The Language
of Mathematics

The Language
of Mathematics

Making the Invisible Visible

KEITH DEVLIN

W. H. FREEMAN AND COMPANY
NEW YORK

Cover Design: Vertigo Design, NYC
Text Design: Diana Blume
Illustrations: Mel Erikson Art Services, Publication Services,
Ian Warpole, Network Graphics

Library of Congress Cataloging-in-Publication Data

Devlin, Keith J.
 The language of mathematics: making the invisible visible / Keith
Devlin.
  p. cm.
 Includes index.
 ISBN 0-7167-3379-X
 1. Mathematics—Popular works. I. Title.
QA93.D4578 1998
510—dc21   98-38019
                                                               CIP

Printed in the United States of America

First printing, 1998

W. H. Freeman and Company
41 Madison Avenue, New York, NY 10010
Houndmills, Basingstoke RG21 6XS, England

Contents

Contents

Preface

This book tries to convey the essence of mathematics, both its historical development and its current breadth. It is not a 'how to' book; it is an 'about' book, which sets out to describe mathematics as a rich and living part of humankind's culture. It is intended for the general reader, and does not assume any mathematical knowledge or ability.

The book grew from an earlier volume in W. H. Freeman's Scientific American Library series, titled *Mathematics: The Science of Patterns*. Written for what is generally termed a 'scientifically literate' audience, that book proved to be one of the most successful in the Scientific American Library series. In conversation with Jonathan Cobb, my editor on that project, the idea arose of a 'spin-off' book aimed at a much wider audience. The new book would not have the high-gloss finish and the masses of full-color artwork and photographs that make the Scientific American Library series so special. Rather, its aim would be to get essentially the same story across in a format that made it accessible to a much wider collection of readers: the story that mathematics is about the identification and study of patterns. (Like its predecessor, this book will show just what counts as a 'pattern' for the mathematician. Suffice it at this point to observe that I am not just talking about wallpaper patterns or patterns on shirts and dresses—though many of those patterns do turn out to have interesting mathematical properties.)

In addition to completely rewriting much of the text to fit a more standard 'popular science book' format, I have also taken advantage of the change of format to add two additional chapters, one on patterns of chance, the other on patterns of the (physical) universe. I had wanted to include both topics in the original book, but there was not enough space in the Scientific American Library format.

Fernando Gouvea, Doris Schattschneider, and Kenneth Millett offered comments on all or parts of the manuscript for the original Scientific American Library book, and their helpful advice has undoubtedly found its way into this new book. Ron Olowin provided helpful feedback on Chapter 8, which along with Chapter 7 is brand new to this book. Susan Moran was the ever-vigilant copy editor for the *Patterns* book. Norma Roche copy-edited this new book.

Historically, almost all leading mathematicians were male, and that is reflected in the almost complete absence of female characters in the book. Those days are, I hope, gone forever. To reflect today's reality, this book uses both 'he' and 'she' interchangeably as the generic third-person pronoun.

What Is Mathematics?

It's not just numbers

What is mathematics? Ask this question of persons chosen at random, and you are likely to receive the answer "Mathematics is the study of numbers." With a bit of prodding as to what *kind* of study they mean, you may be able to induce them to come up with the description "the *science* of numbers." But that is about as far as you will get. And with that, you will have obtained a description of mathematics that ceased to be accurate some two and a half thousand years ago!

Given such a huge misconception, there is scarcely any wonder that your randomly chosen persons are unlikely to realize that research in mathematics is a thriving, worldwide activity, or to accept a suggestion that mathematics permeates, often to a considerable extent, most walks of present-day life and society.

In fact, the answer to the question "What is mathematics?" has changed several times during the course of history.

Up to 500 B.C. or thereabouts, mathematics was indeed the study of numbers. This was the period of Egyptian and Babylonian mathematics. In those civilizations, mathematics consisted almost solely of arithmetic. It was largely utilitarian, and very much of a 'cookbook' nature ("Do such and such to a number and you will get the answer").

The period from around 500 B.C. to A.D. 300 was the era of Greek mathematics. The mathematicians of ancient Greece were primarily concerned with geometry. Indeed, they regarded numbers in a geometric fashion, as measurements of length, and when they discovered that there were lengths to which their numbers did not correspond (irrational lengths), their study of number largely came to a halt. For the Greeks, with their emphasis on geometry, mathematics was the study of number *and shape*.

In fact, it was only with the Greeks that mathematics came into being as an area of study, and ceased being a collection of techniques for measuring, counting, and accounting. The Greeks' interest in mathematics was not just utilitarian; they regarded mathematics as an intellectual pursuit having both aesthetic and religious elements. Thales introduced the idea that the precisely stated assertions of mathematics could be logically proved by a formal argument. This innovation marked the birth of the theorem, now the bedrock of mathematics. For the Greeks, this approach culminated in the publication of Euclid's *Elements*, reputedly the most widely circulated book of all time after the Bible.

Mathematics in motion

There were no major changes in the overall nature of mathematics, and hardly any significant advances within the subject, until the middle of the seventeenth century, when Newton (in England) and Leibniz (in Germany) independently invented the calculus. In essence, the calculus is the study of motion and change. Previous mathematics had been largely restricted to the static issues of counting, measuring, and describing shape. With the introduction of techniques to handle motion and change, mathematicians were able to study the motion of the planets and of falling bodies on earth, the workings of machinery, the flow of liquids, the expansion of gases, physical forces such as magnetism and electricity, flight, the growth of plants and animals, the spread of epidemics, the fluctuation of profits, and so on. After Newton and Leibniz, mathematics became the study of number, shape, *motion, change, and space*.

Most of the initial work involving the calculus was directed toward the study of physics; indeed, many of the great mathematicians of the period are also regarded as physicists. But from about the middle of the eighteenth century there was an increasing interest in the mathematics itself, not just its applications, as mathematicians sought to understand what lay behind the enormous power that the calculus gave to humankind. Here the old Greek tradition of formal proof came back into ascendancy, as a large part of present-day pure mathematics was developed. By the end of the nineteenth century, mathematics had become the study of number, shape, motion, change, and space, *and of the mathematical tools that are used in this study*.

The explosion of mathematical activity that took place in the twentieth century was dramatic. In the year 1900, all the world's mathematical knowledge would have fitted into about eighty books. Today it would take maybe a hundred thousand volumes to contain all known mathematics. This extraordinary growth has not only been a furtherance of previous mathematics; many quite new branches of mathematics have sprung up. In 1900, mathematics could reasonably be regarded as consisting of about twelve distinct subjects: arithmetic, geometry, calculus, and so on. Today, between sixty and seventy distinct categories would be a reasonable figure. Some subjects, such as algebra and topology, have split into various subfields; others, such as complexity theory or dynamical systems theory, are completely new areas of study.

The science of patterns

Given this tremendous growth in mathematical activity, for a while it seemed as though the only simple answer to the question "What is mathematics?" was to say, somewhat fatuously, "It is what mathematicians do for a living." A particular study was classified as mathematics not so much because of *what* was studied but because of *how* it was studied—that is, the methodology used. It was only within the last thirty years or so that a definition of mathematics emerged on which most mathematicians now agree: mathematics is the *science of patterns*. What the mathematician does is examine abstract 'patterns'—numerical patterns, patterns of shape, patterns of motion, patterns of behavior, voting patterns in a population, patterns of repeating chance events, and so on. Those patterns can be either real or imagined, visual or mental, static or dynamic, qualitative or quantitative, purely utilitarian or of little more than recreational interest. They can arise from the world around us, from the depths of space and time, or from the inner workings of the human mind. Different kinds of patterns give rise to different branches of mathematics. For example:

- Arithmetic and number theory study patterns of number and counting.
- Geometry studies patterns of shape.
- Calculus allows us to handle patterns of motion.
- Logic studies patterns of reasoning.
- Probability theory deals with patterns of chance.
- Topology studies patterns of closeness and position.

To convey something of this modern conception of mathematics, this book takes eight general themes, covering patterns of counting, patterns of reasoning and communicating, patterns of motion and change, patterns of shape, patterns of symmetry and regularity, patterns of position, patterns of chance, and the fundamental patterns of the universe. Though this selection leaves out a number of major areas of mathematics, it should provide a good overall sense of what contemporary mathematics is about. The treatment of each theme, while at a purely descriptive level, is not superficial.

One aspect of modern mathematics that is obvious to even the casual observer is the use of abstract notation: algebraic expressions, complicated-looking formulas, and geometric diagrams. The mathematician's reliance on abstract notation is a reflection of the abstract nature of the patterns he studies.

Different aspects of reality require different forms of description. For example, the most appropriate way to study the lay of the land or to describe to someone how to find their way around a strange town is to draw a map. Text is far less appropriate. Analogously, line drawings in the form of blueprints are the most appropriate way to specify the construction of a building. And musical notation is the most appropriate way to convey music, apart from, perhaps, actually playing the piece.

In the case of various kinds of abstract, 'formal' patterns and abstract structures, the most appropriate means of description and analysis is mathematics, using mathematical notation, concepts, and procedures. For instance, the symbolic notation of algebra is the most appropriate means of describing and analyzing the general behavioral properties of addition and multiplication. The commutative law for addition, for example, could be written in English as

When two numbers are added, their order is not important.

However, it is usually written in the symbolic form

$$m + n = n + m.$$

Such is the complexity and the degree of abstraction of the majority of mathematical patterns that to use anything other than symbolic notation would be prohibitively cumbersome. And so the development of mathematics has involved a steady increase in the use of abstract notation.

Symbols of progress

The first systematic use of a recognizably algebraic notation in mathematics seems to have been made by Diophantus, who lived in Alexan-

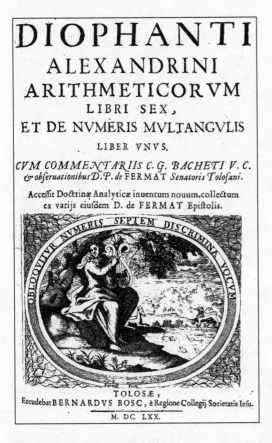

DIOPHANTI
ALEXANDRINI
ARITHMETICORVM
LIBRI SEX,
ET DE NVMERIS MVLTANGVLIS
LIBER VNVS.

CVM COMMENTARIIS C. G. BACHETI V. C.
& obseruationibus D. P. de FERMAT Senatoris Tolosani.

Accessit Doctrinæ Analyticæ inuentum nouum, collectum
ex varijs eiusdem D. de FERMAT Epistolis.

TOLOSÆ,
Excudebat BERNARDVS BOSC, è Regione Collegij Societatis Iesu.
M. DC. LXX.

FIGURE **0.1** The title page of a seventeenth-century Latin translation of Diophantus' classic text *Arithmetic*.

dria sometime around A.D. 250. His treatise *Arithmetic* (Figure 0.1), of which only six of the original thirteen volumes have been preserved, is generally regarded as the first 'algebra textbook'. In particular, Diophantus used special symbols to denote the unknown in an equation and to denote powers of the unknown, and he employed symbols for subtraction and for equality.

These days, mathematics books tend to be awash with symbols, but mathematical notation no more *is* mathematics than musical notation *is* music (see Figure 0.2). A page of sheet music *represents* a piece of music; the music itself is what you get when the notes on the page are sung or performed on a musical instrument. It is in its performance that the music comes alive and becomes part of our experience; the music exists not on the printed page, but in our minds. The same is true for mathematics; the symbols on a page are just a representation of the mathematics. When read by a competent performer (in this case, someone trained in mathematics), the symbols on the printed page come alive—the mathematics lives and breathes in the mind of the reader like some abstract symphony.

FIGURE **0.2** Like mathematics, music has an abstract notation, used to represent abstract structures.

Given the strong similarity between mathematics and music, both of which have their own highly abstract notations and are governed by their own structural rules, it is hardly surprising that many (perhaps most) mathematicians also have some musical talent.

In fact, for most of the two and a half thousand years of Western civilization, starting with the ancient Greeks, mathematics and music were regarded as two sides of the same coin: both were thought to provide insights into the order of the universe. It was only with the rise of the scientific method in the seventeenth century that the two started to go their separate ways.

For all their historical connections, however, there was, until recently, one very obvious difference between mathematics and music. Though only someone well trained in music can read a musical score and hear the music in her head, if that same piece of music is performed by a competent musician, anyone with the sense of hearing can appreciate the result. It requires no musical training to experience and enjoy music when it is performed.

For most of its history, however, the only way to appreciate mathematics was to learn how to 'sight-read' the symbols. Though the structures and patterns of mathematics reflect the structure of, and resonate in, the human mind every bit as much as do the structures and patterns of music, human beings have developed no mathematical equivalent of a pair of ears. Mathematics can be 'seen' only with the 'eyes of the mind'. It is as if we had no sense of hearing, so that only someone able to sight-read musical notation would be able to appreciate the patterns and harmonies of music.

In recent years, however, the development of computer and video technologies has to some extent made mathematics accessible to the untrained. In the hands of a skilled user, the computer can be used to 'perform' mathematics, and the result can be displayed in a visual form on the screen for all to see. Though only a relatively small part of mathematics lends itself to such visual 'performance', it is now possible to convey to the layperson at least something of the beauty and the harmony that the mathematician 'sees' and experiences when she does mathematics.

When seeing is discovering

Sometimes computer graphics can be of significant use to the mathematician as well as providing the layperson with a glimpse of the inner world of mathematics. For instance, the study of complex dynamical systems was begun in the 1920s by the French mathematicians Pierre Fatou and Gaston Julia, but it was not until the late 1970s and early 1980s that the rapidly developing technology of computer graphics enabled Benoit Mandelbrot and other mathematicians to see some of the structures Fatou and Julia had been working with. The strikingly beautiful pictures that emerged from this study have since become something of an art form in their own right. In honor of one of the two pioneers of the subject, certain of these structures are now called Julia sets (see Figure 0.3).

FIGURE 0.3 A Julia set.

Another example of the use of computer graphics leading to a deep discovery in mathematics occurred in 1983, when mathematicians David Hoffman and William Meeks III discovered a brand new minimal surface (see Plate 1). A *minimal surface* is the mathematical equivalent of an infinite soap film. Real soap films stretched across a frame always form a surface that occupies the minimal possible area. The mathematician considers abstract analogues of soap films that stretch out to infinity. Such surfaces have been studied for over two hundred years, but, until Hoffman and Meeks made their discovery, only three such surfaces were known. Today, as a result of computer visualization techniques, mathematicians have discovered many such surfaces. Much of what is known about minimal surfaces has been established by more traditional mathematical techniques, involving lots of algebra and calculus. But, as Hoffman and Meeks showed, computer graphics can provide the mathematician with the intuition needed to find the right combination of those traditional techniques.

Without its algebraic symbols, large parts of mathematics simply would not exist. Indeed, the issue is a deep one, having to do with human cognitive abilities. The recognition of abstract concepts and the development of an appropriate language to represent them are really two sides of the same coin.

The use of a symbol such as a letter, a word, or a picture to denote an abstract entity goes hand in hand with the recognition of that entity *as an entity*. The use of the numeral '7' to denote the number 7 requires that the number 7 be recognized as an entity; the use of the letter m to denote an arbitrary whole number requires that the *concept* of a whole number be recognized. Having the symbol makes it possible to think about and manipulate the concept.

This linguistic aspect of mathematics is often overlooked, especially in our modern culture, with its emphasis on the procedural, computational aspects of mathematics. Indeed, one often hears the complaint that mathematics would be much easier if it weren't for all that abstract notation, which is rather like saying that Shakespeare would be much easier to understand if it were written in simpler language.

Sadly, the level of abstraction in mathematics, and the consequent need for notation that can cope with that abstraction, means that many, perhaps most, parts of mathematics will remain forever hidden from the nonmathematician; and even the more accessible parts—the parts described in books such as this one—may be at best dimly perceived, with much of their inner beauty locked away from view. Still, that does not excuse those of us who do seem to have been blessed with an ability to appreciate that inner beauty from trying to communicate to others some sense of what it is we experience—some sense of the simplicity, the precision, the purity, and the elegance that give the patterns of mathematics their aesthetic value.

The hidden beauty in the symbols

In his 1940 book *A Mathematician's Apology*, the accomplished English mathematician G. H. Hardy wrote:

> The mathematician's patterns, like the painter's or the poet's, must be *beautiful*, the ideas, like the colors or the words, must fit together in a harmonious way. Beauty is the first test; there is no permanent place in the world for ugly mathematics. . . . It may be very hard to *define* mathematical beauty, but that is just as true of beauty of any kind—we may not know quite what we mean by a beautiful poem, but that does not prevent us from recognizing one when we read it.

The beauty to which Hardy was referring is, in many cases, a highly abstract, *inner* beauty, a beauty of abstract form and logical structure, a beauty that can be observed, and appreciated, only by those sufficiently well trained in the discipline. It is a beauty "cold and austere," according to Bertrand Russell, the famous English mathematician and philosopher, who wrote, in his 1918 book *Mysticism and Logic:*

> Mathematics, rightly viewed, possesses not only truth, but supreme beauty—a beauty cold and austere, like that of sculpture, without appeal to any part of our weaker nature, without the gorgeous trappings of painting or music, yet sublimely pure, and capable of a stern perfection such as only the greatest art can show.

Mathematics, the science of patterns, is a way of looking at the world, both the physical, biological, and sociological world we inhabit and the inner world of our minds and thoughts. Mathematics' greatest success has undoubtedly been in the physical domain, where the subject is rightly referred to as both the queen and the servant of the (natural) sciences. Yet, as an entirely human creation, the study of mathematics is ultimately a study of humanity itself. For none of the entities that form the substrate of mathematics exist in the physical world; the numbers, the points, the lines and planes, the surfaces, the geometric figures, the functions, and so forth are pure abstractions that exist only in humanity's collective mind. The absolute certainty of a mathematical proof and the indefinitely enduring nature of mathematical truth are reflections of the deep and fundamental status of the mathematician's patterns in both the human mind and the physical world.

In an age when the study of the heavens dominated scientific thought, Galileo said,

> The great book of nature can be read only by those who know the language in which it was written. And this language is mathematics.

Striking a similar note in a much later era, when the study of the inner workings of the atom had occupied the minds of many scientists for a generation, the Cambridge physicist John Polkinhorne wrote, in 1986,

> Mathematics is the abstract key which turns the lock of the physical universe.

In today's age, dominated by information, communication, and computation, mathematics is finding new locks to turn. There is scarcely any aspect of our lives that is not affected, to a greater or lesser extent, by mathematics, for abstract patterns are the very essence of thought, of communication, of computation, of society, and of life itself.

Making the invisible visible

We have answered the question "What is mathematics?" with the slogan "Mathematics is the science of patterns." There is another fundamental question about mathematics that can also be answered by a catchy phrase: "What does mathematics do?" By this I mean, what exactly does mathematics give you when you apply it to the study of some phenomenon? The answer is "Mathematics makes the invisible visible."

Let me give you some examples of what I mean by this answer.

Without mathematics, there is no way you can understand what keeps a jumbo jet in the air. As we all know, large metal objects don't stay above the ground without something to support them. But when you look at a jet aircraft flying overhead, you can't see anything holding it up. It takes mathematics to 'see' what keeps an airplane aloft. In this case, what lets you 'see' the invisible is an equation discovered by the mathematician Daniel Bernoulli early in the eighteenth century.

While I'm on the subject of flying, what is it that causes objects other than aircraft to fall to the ground when we release them? "Gravity," you answer. But that's just giving it a name; it doesn't help us to understand it. It's still invisible. We might as well call it 'magic'. To understand gravity, you have to 'see' it. That's exactly what Newton did with his equations of motion and mechanics in the seventeenth century. Newton's mathematics enabled us to 'see' the invisible forces that keep the earth rotating around the sun and cause an apple to fall from a tree onto the ground.

Both Bernoulli's equation and Newton's equations use calculus. Calculus works by making visible the infinitesimally small. That's another example of making the invisible visible.

Here's another: Two thousand years before we could send spacecraft into outer space to provide us with pictures of our planet, the Greek mathematician Eratosthenes used mathematics to show that the earth was round. Indeed, he calculated its diameter, and hence its curvature, with 99 percent accuracy.

Today, we may be close to repeating Eratosthenes' feat by discovering whether the universe is curved. Using mathematics and powerful telescopes, we can 'see' into the outer reaches of the universe. According to some astronomers, we will soon see far enough to be able to detect any curvature in space, and to measure any curvature that we find.

Knowing the curvature of space, we can then use mathematics to see into the future to the day the universe comes to an end. Using mathematics, we have already been able to see into the distant past, making visible the otherwise invisible moments when the universe was first created in what we call the Big Bang.

Coming back to earth at the present time, how do you 'see' what makes pictures and sounds of a football game miraculously appear on a television screen on the other side of town? One answer is that the pictures and sounds are transmitted by radio waves—a special case of what we call electromagnetic radiation. But, as with gravity, that answer just gives the phenomenon a name; it doesn't help us to 'see' it. In order to 'see' radio waves, you have to use mathematics. Maxwell's equations, discovered in the nineteenth century, make visible to us the otherwise invisible radio waves.

Here are some human patterns we can 'see' through mathematics:

- Aristotle used mathematics to try to 'see' the invisible patterns of sound that we recognize as music.
- He also used mathematics to try to describe the invisible structure of a dramatic performance.
- In the 1950s, the linguist Noam Chomsky used mathematics to 'see' and describe the invisible, abstract patterns of words that we recognize as grammatical sentences. He thereby turned linguistics from a fairly obscure branch of anthropology into a thriving mathematical science.

Finally, using mathematics, we are able to look into the future:

- Probability theory and mathematical statistics let us predict the outcomes of elections, often with remarkable accuracy.

- We use calculus to predict tomorrow's weather.
- Market analysts use various mathematical theories to try to predict the behavior of the stock market.
- Insurance companies use statistics and probability theory to predict the likelihood of an accident during the coming year, and set their premiums accordingly.

When it comes to looking into the future, mathematics allows us to make visible another invisible—that which has not yet happened. In that case, our mathematical vision is not perfect. Our predictions are sometimes wrong. But without mathematics, we cannot see into the future even poorly.

The invisible universe

Today, we live in a technological society. There are increasingly few places on the face of the earth where, when we look around us toward the horizon, we do not see products of our technology: tall buildings, bridges, power lines, telephone cables, cars on roads, aircraft in the sky. Where communication once required physical proximity, today much of our communication is mediated by mathematics, transmitted in digitized form along wires or optical fibers, or through the ether. Computers— machines that perform mathematics—are not only on our desktops, they are in everything from microwave ovens to automobiles and from children's toys to pacemakers for those with heart problems. Mathematics— in the form of statistics—is used to decide what food we will eat, what products we will buy, what television programs we will be able to see, and which politicians we will be able to vote for. Just as society burned fossil fuels to drive the engines of the industrial age, in today's information age, the principal fuel we burn is mathematics.

And yet, as the role of mathematics has grown more and more significant over the past half century, it has become more and more hidden from view, forming an invisible universe that supports much of our lives. Just as our every action is governed by the invisible forces of nature (such as gravity), we now live in the invisible universe created by mathematics, subject to invisible mathematical laws.

This book will take you on a tour of that invisible universe. It will show you how we can use mathematics to see some of its invisible structure. You may find some of the sights you encounter on this tour strange and unfamiliar, like those of a far-off land. But for all its unfamiliarity, this is not a distant universe through which we will be traveling; it's a universe we all live in.

CHAPTER 1

Why Numbers Count

You can count on them

Numbers—that is to say, whole numbers—arise from the recognition of patterns in the world around us: the pattern of 'oneness', the pattern of 'twoness', the pattern of 'threeness', and so on. To recognize the pattern that we call 'threeness' is to recognize what it is that a collection of three apples, three children, three footballs, and three rocks have in common. "Can you see a pattern?" a parent might ask a small child, showing her various collections of objects—three apples, three shoes, three gloves, and three toy trucks. The counting numbers 1, 2, 3, . . . are a way of capturing and describing those patterns. The patterns captured by numbers are abstract, and so are the numbers used to describe them.

Having arrived at the number concept as an abstraction of certain patterns in the world around us, another pattern arises at once, a mathematical pattern of the numbers. The numbers are ordered 1, 2, 3, . . . , each succeeding number being greater by 1 than its predecessor.

There are still deeper patterns of number to be examined by the mathematician, patterns of evenness and oddness, of being prime or composite, of being a perfect square, of satisfying various equations, and so forth. The study of number patterns of this form is known as *number theory*.

These days, children do it before they're five

At the age of five or less, the typical child in an educated, Western culture makes a cognitive leap that took humankind many thousands of years to achieve: the child acquires the concept of number. He or she comes to realize that there is something common to a collection of, say, five apples, five oranges, five children, five cookies, a rock group of five members, and so on. That common something, 'fiveness', is somehow captured or encapsulated by the number 5, an abstract entity that the child will never see, hear, feel, smell, or taste, but which will have a definite existence for the rest of his or her life. Indeed, such is the role numbers play in everyday life that for most people, the ordinary counting numbers 1, 2, 3, ... are more real, more concrete, and certainly more familiar than Mount Everest or the Taj Mahal.

The conceptual creation of the counting numbers marks the final step in the process of recognizing the *pattern* of 'number of members of a given collection'. This pattern is completely abstract—indeed, so abstract that it is virtually impossible to talk about it except in terms of the abstract numbers themselves. Try explaining what is meant by a collection of twenty-five objects without referring to the *number* 25. (With a very small collection, you can make use of your fingers: a collection of five objects can be explained by holding up the fingers of one hand and saying "This many.")

The acceptance of abstraction does not come easily to the human mind. Given the choice, people prefer the concrete over the abstract. Indeed, work in psychology and anthropology indicates that a facility with abstraction seems to be something we are not born with, but acquire, often with great difficulty, as part of our intellectual development.

For instance, according to the work of cognitive psychologist Jean Piaget, the abstract concept of volume is not innate, but is learned at an early age. Young children are not able to recognize that a tall, thin glass and a short, stout one can contain the same volume of liquid, even if they see the one poured into the other. For a considerable time, they will maintain that the quantity of liquid changes, that the tall glass contains more than the short one.

The concept of abstract number also appears to be learned. Small children seem to acquire this concept after they have learned to count. Evidence that the concept of number is not innate comes from the study of cultures that have evolved in isolation from modern society.

For instance, when a member of the Vedda tribe of Sri Lanka wants to count coconuts, he collects a heap of sticks and assigns one to each coconut. Each time he adds a new stick, he says, "That is one." But if asked to say how many coconuts he possesses, he simply points to the pile of sticks and says, "That many." The tribesman thus has a type of counting system, but far from using abstract numbers, he 'counts' in terms of decidedly concrete sticks.

The Vedda tribesman employs a system of counting that dates back to very early times, that of using one collection of objects—say, sticks or pebbles—to 'count' the members of another collection, by pairing off the sticks or pebbles with the objects to be 'counted'.

A token advance

The earliest known human-made artifacts believed to be connected with counting are notched bones, some of which date back to around 35,000 B.C. At least in some cases, these bones seem to have been used as lunar calendars, with each notch representing one sighting of the moon. Similar instances of counting by means of a one-to-one correspondence appear again and again in preliterate societies: pebbles and shells were used in the census in early African kingdoms, and cacao beans and kernels of maize, wheat, and rice were used as counters in the New World.

Of course, any such system suffers from an obvious lack of specificity. A collection of notches, pebbles, or shells indicates a quantity, but not the kinds of items being quantified, and hence cannot serve as a means of storing information for long periods. The first known enumeration system that solved this problem was devised in what is now the Middle East, in the region known as the Fertile Crescent, stretching from present-day Syria to Iran.

During the 1970s and early 1980s, anthropologist Denise Schmandt-Besserat of the University of Texas at Austin carried out a detailed study of clay artifacts found in archaeological digs at various locations in the Middle East. At every site, among the usual assortment of clay pots, bricks, and figurines, Schmandt-Besserat noticed the presence of collections of small, carefully crafted clay shapes, each measuring between 1 and 3 centimeters across: spheres, disks, cones, tetrahedrons, ovoids, cylinders, triangles, rectangles, and the like (see Figure 1.1). The earliest such objects dated back to around 8000 B.C., some time after people had started to develop agriculture and would have first needed to plan harvests and lay down stores of grain for later use.

An organized agriculture requires a means of keeping track of a person's stock, and a means to plan and to barter. The clay shapes examined by Schmandt-Besserat appear to have been developed to fulfill this role, with the various shapes being used as tokens to represent the kind of object being counted. For example, there is evidence that a cylinder

FIGURE 1.1 Clay artifacts like these found in Susa, Iran, were used for accounting in systems of organized agriculture in the Fertile Crescent. *Top:* Complex tokens representing (*top row, left to right*) 1 sheep, 1 unit of a particular oil (?), 1 unit of metal, 1 type of garment, and (*bottom row*) 1 garment of a second type, an unknown commodity, and 1 measure of honey. All ca. 3300 B.C. *Middle:* An envelope and its contents of tokens and the corresponding markings, ca. 3300 B.C. *Bottom:* An impressed tablet featuring an account of grain, ca. 3100 B.C.

stood for an animal, cones and spheres stood for two common measures of grain (approximately a peck and a bushel, respectively), and a circular disk stood for a flock. In addition to providing a convenient, physical record of a person's holdings, the clay shapes could be used in planning and bartering, by means of physical manipulation of the tokens.

By 6000 B.C., the use of clay tokens had spread throughout the region. The nature of the clay tokens remained largely unchanged until around 3000 B.C., when the increasingly more complex societal structure of the Sumerians—characterized by the growth of cities, the rise of the Sumerian temple institution, and the development of organized government—led to the development of more elaborate forms of tokens. These newer tokens had a greater variety of shapes, including rhomboids, bent coils, and parabolas, and were imprinted with markings. Whereas the plain tokens continued to be used for agricultural accounting, these more complex tokens appear to have been introduced to represent manufactured objects such as garments, metalworks, jars of oil, and loaves of bread.

The stage was set for the next major step toward the development of abstract numbers. During the period 3300 to 3250 B.C., as state bureaucracy grew, two means of storing clay tokens became common. The more elaborate, marked tokens were perforated and strung together on a string attached to an oblong clay frame, and the frame was marked to indicate the identity of the account in question. The plain tokens were stored in clay containers, hollow balls some 5 to 7 centimeters in diameter, and the containers were marked to show the account. Both the strings of tokens and the sealed clay envelopes of tokens thus served as accounts or contracts.

Of course, one obvious drawback of a sealed clay envelope is that the seal has to be broken open in order to examine the contents. So Sumerian accountants developed the practice of impressing the tokens on the soft exteriors of the envelopes before enclosing them, thereby leaving a visible exterior record of the contents.

But with the contents of the envelope recorded on the exterior, the tokens themselves became largely superfluous: all the requisite information was stored in the envelope's outer markings. The tokens themselves could be discarded, which is precisely what happened after a few generations. The result was the birth of the clay tablet, on which impressed marks, and those marks alone, served to record the data previously represented by the tokens. In present-day terminology, we would say that the Sumerian accountants had replaced the physical counting devices with written *numerals*.

From a cognitive viewpoint, it is interesting that the Sumerians did not immediately advance from using physical tokens sealed in a marked envelope to using markings on a single tablet. For some time, the marked clay envelopes redundantly contained the actual tokens depicted by their outer markings. The tokens were regarded as representing the quantity of grain, the number of sheep, or whatever; the envelope's outer markings were regarded as representing not the real-world quantity, but the tokens in the envelope. That it took so long to recognize the redundancy of the tokens suggests that going from physical tokens to an abstract representation was a considerable cognitive development.

Of course, the adoption of a symbolic representation of an amount of grain does not in itself amount to the explicit recognition of the number concept in the sense familiar today, in which numbers are regarded as 'things', as 'abstract objects'. Exactly when humankind achieved that feat is hard to say, just as it is not easy to pinpoint the moment when a small child makes a similar cognitive advance. What is certain is that, once the clay tokens had been abandoned, the functioning of Sumerian society relied on the notions of 'oneness', 'twoness', 'threeness', and so on, since that is what the markings on their tablets denoted.

Symbolic progress

Having some kind of written numbering system, and using that system to count, as the Sumerians did, is one thing; recognizing a concept of number and investigating the properties of numbers—developing a *science* of numbers—is quite another. This latter development came much later, when people first began to carry out intellectual investigations of the kind that we would now classify as science.

As an illustration of the distinction between the *use* of a mathematical device and the explicit recognition of the entities involved in that device, take the familiar observation that order is not important when a pair of counting numbers are added or multiplied. (From now on I shall generally refer to counting numbers by the present-day term *natural numbers*.) Using modern algebraic terminology, this principle can be expressed in a simple, readable fashion by the two commutative laws:

$$m + n = n + m, \qquad m \times n = n \times m.$$

In each of these two identities, the symbols m and n are intended to denote *any* two natural numbers. Using these symbols is quite different from writing down a particular instance of these laws, for example:

$$3 + 8 = 8 + 3, \qquad 3 \times 8 = 8 \times 3.$$

The second case is an observation about the addition and multiplication of two particular numbers. It requires our having the ability to handle individual abstract numbers, at the very least the abstract numbers 3 and 8, and is typical of the kind of observation that was made by the early Egyptians and Babylonians. But it does not require a well-developed *concept* of abstract number, as do the commutative laws.

By around 2000 B.C., both the Egyptians and the Babylonians had developed primitive numeral systems and made various geometric observations concerning triangles, pyramids, and the like. Certainly they 'knew' that addition and multiplication were commutative, in the sense that they were familiar with these two patterns of behavior, and undoubtedly made frequent use of commutativity in their daily calculations. But in their writings, when describing how to perform a particular kind of computation, they did not use algebraic symbols such as *m* and *n*. Instead, they always referred to *particular* numbers, although it is clear that in many cases the particular numbers chosen were presented purely as examples, and could be freely replaced by any other numbers.

For example, in the so-called Moscow Papyrus, an Egyptian document written in 1850 B.C., appear the following instructions for computing the volume of a certain truncated square pyramid (one with its top chopped off by a plane parallel to the base—see Figure 1.2):

> If you are told: a truncated pyramid of 6 for the vertical height by 4 on the base by 2 on the top. You are to square this 4, result 16. You are to double 4, result 8. You are to square 2, result 4. You are to add the 16, the 8, and the 4, result 28. You are to take a third of 6, result 2. You are to take 28 twice, result 56. See, it is 56. You will find it right.

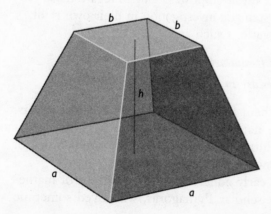

FIGURE **1.2** A truncated square pyramid.

Though these instructions are given in terms of particular dimensions, they clearly make sense as a set of instructions only if the reader is free to replace these numbers with any other appropriate values. In modern notation, the result would be expressed by means of an algebraic formula: if the truncated pyramid has a base of sides equal to a, a top of sides equal to b, and a height h, then its volume is given by the formula

$$V = \frac{1}{3}h(a^2 + ab + b^2).$$

Being aware of, and utilizing, a certain pattern is not the same as formalizing that pattern and subjecting it to a scientific analysis. The commutative laws, for example, express certain patterns in the way the natural numbers behave under addition and multiplication, and moreover, they express these patterns in an explicit fashion. By formulating the laws using algebraic indeterminates such as m and n, entities that denote *arbitrary* natural numbers, we place the focus very definitely on the pattern, not the addition or the multiplication itself.

The general concept of abstract number was not recognized, nor were behavioral rules such as those concerning addition and multiplication formulated, until the era of Greek mathematics began around 600 B.C.

For a long time it was all Greek

It is not possible to say exactly when abstract mathematics first appeared, but if a time and place had to be set, it would most likely be the sixth century B.C. in Greece, when Thales of Miletus carried out his investigations of geometry. Thales' travels as a merchant undoubtedly exposed him to the known geometric ideas involved in measurement, but it was apparently not until his own contributions that any attempt was made to regard those geometric ideas as a subject for systematic investigation in their own right.

Thales took known observations such as:

A circle is bisected by any of its diameters.

The sides of similar triangles are in proportion.

and showed how to deduce them from other, supposedly more 'basic', observations concerning the nature of length and area. The idea of mathematical proof thereby introduced was to become the bedrock of much of mathematics to follow.

One of the most famous early adherents to the concept of mathematical proof was the Greek scholar Pythagoras, who lived sometime around 570 to 500 B.C. Few details of his life are known, since both he and his followers shrouded themselves in mystery, regarding their mathematical studies as something of a black art. He is believed to have been born between 580 and 560 B.C. on the Aegean island of Samos, and to have studied in both Egypt and Babylonia. After several years of wandering, he appears to have settled in Croton, a prosperous Greek settlement in southern Italy. The school he founded there concentrated on the study of *arithmetica* (number theory), *harmonia* (music), *geometria* (geometry), and *astrologia* (astronomy), a fourfold division of knowledge that in the Middle Ages became known as the *quadrivium*. Together with the *trivium* of logic, grammar, and rhetoric, the *quadrivium* made up the seven 'liberal arts' that were regarded as constituting a necessary course of study for an educated person.

Mixed up with the Pythagoreans' philosophical speculations and mystical numerology was some genuinely rigorous mathematics, including the famous Pythagorean theorem. Illustrated in Figure 1.3, the theorem states that for any right-angled triangle, the square of the length of the hypotenuse is equal to the sum of the squares of the lengths of the other two sides. This theorem is remarkable on two counts. First, the Pythagoreans were able to discern the relationship between the squares of the sides, observing that there was a regular pattern that was exhibited by *all* right-angled triangles. Second, they were able to come up with a rigorous proof that the pattern they had observed did indeed hold for all such triangles.

The abstract patterns of principal interest to the Greek mathematicians were geometric ones—patterns of shape, angle, length, and area. Indeed, apart from the natural numbers, the Greek notion of number was essentially based on geometry, with numbers being thought of as measurements of length and area. All their results concerning angles,

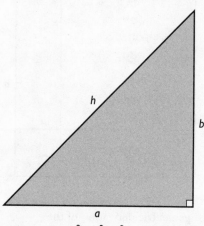

$$h^2 = a^2 + b^2$$

FIGURE **1.3** The Pythagorean theorem relates the length of the hypotenuse (h) of a right-angled triangle to the lengths of the other two sides (a and b).

lengths, and areas—results that would nowadays be expressed in terms of whole numbers and fractions—were given by the Greeks in the form of comparisons of one angle, length, or area with another. It was this concentration on *ratios* that gave rise to the modern term *rational number* for a number that can be expressed as the quotient of one whole number by another.

The Greeks discovered various algebraic identities familiar to present-day students of mathematics, such as:

$$(a + b)^2 = a^2 + 2ab + b^2,$$
$$(a - b)^2 = a^2 - 2ab + b^2.$$

Again, these were thought of in geometric terms, as observations about adding and subtracting areas. For example, in Euclid's *Elements* (of which more follows), the first of these algebraic identities is stated as follows:

> Proposition II.4. If a straight line be cut at random, the square on the whole is equal to the squares on the segments and twice the rectangle contained by the segments.

This proposition is illustrated by the left-hand diagram in Figure 1.4.

In the diagram, the area of the large square = $(a + b)^2$ = the area of square A plus the area of square B plus the area of rectangle C plus the area of rectangle $D = a^2 + b^2 + ab + ab = a^2 + 2ab + b^2$. The second identity is derived from the diagram on the right in Figure 1.4, in which

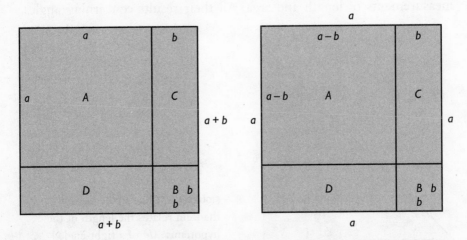

FIGURE 1.4 These diagrams show the Greeks' geometric derivation of the algebraic identities for $(a + b)^2$ (*left*) and $(a - b)^2$ (*right*).

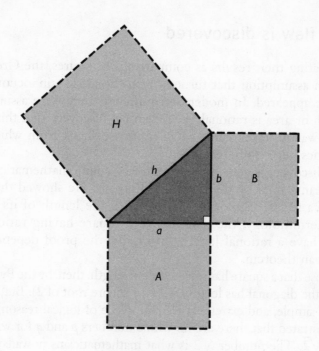

Area H = area A + area B

FIGURE 1.5 The Greeks understood the Pythagorean theorem geometrically, in terms of the areas of square figures drawn along the three sides of a right-angled triangle. In terms of areas, the theorem says that $H = A + B$.

the sides are labeled differently. In this figure, the area of square $A = (a - b)^2$ = the area of the large square minus the area of the rectangle comprising regions C and B minus the area of rectangles comprising D and B plus the area of square B (added on since this area has been subtracted twice, once as part of each rectangle) = $a^2 - ab - ab + b^2$.

Incidentally, the Greek number system did not include negative numbers. Indeed, negative numbers did not come into widespread use until as recently as the eighteenth century.

The Pythagorean theorem can nowadays be expressed by means of the algebraic identity

$$h^2 = a^2 + b^2,$$

where h is the length of the hypotenuse of a given right-angled triangle and a, b are the lengths of the other two sides. The Greeks, however, understood and proved the theorem in purely geometric terms, as a result about the areas of square figures drawn along the three sides of a given triangle, as illustrated in Figure 1.5.

23

A fatal flaw is discovered

In formulating their results as comparisons of figures, the Greeks were making an assumption that turned out to be far less innocuous than it must have appeared. In modern terminology, they were assuming that any length or area is rational. The eventual discovery that this particular belief was mistaken came as an immense shock from which Greek mathematics never fully recovered.

The discovery is generally credited to a young mathematician in the Pythagorean school by the name of Hippasus. He showed that the diagonal of a square cannot be compared to the length of its sides—in modern terminology, the diagonal of a square having rational sides does not have a rational length. Ironically, the proof depends on the Pythagorean theorem.

Suppose that a square has sides 1 unit in length; then, by the Pythagorean theorem, the diagonal has length $\sqrt{2}$ (the square root of 2). But, by means of a fairly simple, and extremely elegant, piece of logical reasoning, it can be demonstrated that there are no whole numbers p and q for which p/q is equal to $\sqrt{2}$. The number $\sqrt{2}$ is what mathematicians nowadays refer to as an *irrational number*. Here is the simple, yet elegant, proof.

Start out by supposing that, contrary to what I said above, there were natural numbers p and q for which $p/q = \sqrt{2}$. If p and q have any common factors, we can cancel them out, so we may as well assume this has already been done, and that p and q have no common factors.

Squaring the identity $\sqrt{2} = p/q$ gives $2 = p^2/q^2$, which rearranges to give $p^2 = 2q^2$. This equation tells us that p^2 is an even number. Now, the square of any even number is even, and the square of any odd number is odd. So, as p^2 is even, it must be the case that p is even. Consequently, p is of the form $p = 2r$, for some natural number r. Substituting $p = 2r$ into the identity $p^2 = 2q^2$ gives $4r^2 = 2q^2$, which simplifies to $2r^2 = q^2$. This equation tells us that q^2 is an even number. It follows, as in the case of p, that q is itself even.

But now we have shown that both p and q are even, which is contrary our assumption at the start that p and q have no common factors. This contradiction implies that our original assumption that such natural numbers as p and q exist must have been false. In other words, there are no such p and q.

And that's the proof!

Such is the power of proof in mathematics that there was no question of ignoring the new result, even though some popular accounts maintain that Hippasus was thrown from a ship and drowned in order to prevent the terrible news from breaking out.

Unfortunately, instead of provoking a search for a system of numbers richer than the rationals—a step that was to come only much later in history, with the development of the 'real numbers'—Hippasus' discovery was regarded as a fundamental impasse. From then on, the Greeks tended to regard the study of number as distinct from the study of geometry, and their most impressive discoveries about numbers were largely concerned not with measurements of length or area, but with the natural numbers. The first systematic investigation of the natural numbers is generally identified with Euclid, who lived sometime in the period from 350 to 300 B.C.

Here's looking at Euclid

In between the time of Thales and Pythagoras and the arrival of Euclid on the scene, Greek mathematics made considerable advances with the work of Socrates, Plato, Aristotle, and Eudoxus. It was at the Athens Academy, founded by Plato, that Eudoxus worked. There he developed, among other things, a 'theory of proportions' that enabled the Greeks to circumvent, in part, some of the problems created by Hippasus' discovery. Euclid, too, is reputed to have studied at Plato's Academy before settling in the new intellectual center of Alexandria, sometime around 330 B.C.

While working in Alexandria at the great Library, the forerunner of today's universities, Euclid produced his mammoth, thirteen-volume work *Elements*. It was a compendium of practically all of Greek mathematics up to the time, containing some 465 propositions from plane and solid geometry and from number theory. Though some of the results were Euclid's own, for the most part his great contribution was the systematic manner in which the mathematics was presented.

Over the centuries since it was written, more than two thousand editions of *Elements* have been published, and though it contains a number of logical flaws, it remains an excellent example of what we call the mathematical method, in which we commence with a precise statement of the basic assumptions and thereafter accept as facts only those results proved from those assumptions.

Books I to VI of *Elements* concentrate on plane geometry and Books XI to XIII deal with solid geometry, both of which are covered in Chapter 4 of this book. Book X presents an investigation of so-called 'incommensurable magnitudes'. Translated into modern terminology, this volume would be a study of the irrational numbers. It is in Books VII to IX that Euclid presents his treatment of what is now known as number theory, the study of the natural numbers. An obvious pattern exhibited by the natural numbers is that they are ordered one after the other. Number theory examines deeper mathematical patterns found in the natural numbers.

Numbers in prime condition

Euclid begins Book VII of *Elements* with a list of some twenty-two basic definitions, among which are the following: An *even* number is one that is divisible into two equal whole-number parts, and an *odd* number is one that is not. Somewhat more significantly, a *prime* number is (in modern terminology) one that has no whole-number divisors other than 1 and the number itself. For example, among the numbers 1 to 20, the numbers 2, 3, 5, 7, 11, 13, 17, and 19 are primes. A number greater than 1 that is not prime is said to be *composite*. Thus, 4, 6, 8, 9, 10, 12, 14, 15, 16, 18, and 20 are the composite numbers in the range 1 to 20.

Among the fundamental results Euclid proved about the primes are the following:

- If a prime number p divides a product mn, then p divides at least one of the two numbers m, n.
- Every natural number is either prime or else can be expressed as a product of primes in a way that is unique apart from the order in which they are written.
- There are infinitely many primes.

The second of these results is of such importance that it is generally referred to as the *fundamental theorem of arithmetic*. Taken together, the first two results tell us that the primes are very much like the physicist's atoms, in that they are the fundamental building blocks out of which all other natural numbers can be built, in this case through the process of multiplication. For example:

$$328,152 = 2 \times 2 \times 2 \times 3 \times 11 \times 11 \times 113.$$

Each of the numbers 2, 3, 11, 113 is prime; they are called the *prime factors* of 328,152. The product $2 \times 2 \times 2 \times 3 \times 11 \times 11 \times 113$ is called the *prime decomposition* of 328,152. As with atomic structure, a knowledge of the prime decomposition of a given number can enable the mathematician to say a great deal about the mathematical properties of that number.

The third result, the infinitude of the primes, might come as a surprise to anyone who has spent time enumerating prime numbers. Though primes seem to be in great abundance among the first hundred or so natural numbers, they start to thin out as you proceed up through the numbers, and it is not at all clear from the observational evidence whether or not they eventually peter out altogether. For instance, there are eight primes between 2 and 20, but only four between 102 and 120. Going further, of the hundred numbers between 2,101 and 2,200, only ten are prime, and of the hundred between 10,000,001 and 10,000,100, only two are prime.

Prime order

One way to get a precise sense of how the primes thin out is to look at what is called the *prime density function*, which gives the proportion of numbers below a given number that are primes. To obtain the prime density for a given number N, you take the number of primes less than N, call it $\pi(N)$, and divide it by N. In the case of $N = 100$, for example, the answer is 0.168, which tells you that about 1 in 6 of the numbers below 100 are primes. For $N = 1,000,000$, however, the proportion drops to 0.078, which is about 1 in 13, and for $N = 100,000,000$ it is 0.058, or about 1 in 17. As N increases, this fall continues.

N	$\pi(N)$	$\pi(N)/N$
1,000	168	0.168
10,000	1,229	0.123
100,000	9,592	0.095
1,000,000	78,498	0.078
10,000,000	664,5790	0.066
100,000,000	5,761,455	0.058

But for all of this steady fall in the ratio $\pi(N)/N$, the primes never peter out completely. Euclid's proof of this fact remains to this day a marvelous example of logical elegance. Here it is.

The idea is to demonstrate that if you start to list the primes as a sequence $p_1, p_2, p_3, \ldots$, this list continues forever. To prove this, you show that if you have listed all the primes up to some prime p_n, then you can always find another prime to add to the list: the list never stops.

Euclid's ingenious idea was to look at the number

$$P = p_1 \times p_2 \times \ldots \times p_n + 1,$$

where $p_1, \ldots, p_n$ are all the primes enumerated so far. If P happens to be prime, then P is a prime bigger than all the primes $p_1, \ldots, p_n$, so the list may be continued. (P might not be the *next* prime after p_n, in which case you will not take P to be p_{n+1}. But if P is prime, then you know for sure that *there is* a next prime beyond p_n.)

On the other hand, if P is not prime, then P must be evenly divisible by a prime. But none of the primes $p_1, \ldots, p_n$ divides P; if you try to carry out such a division, you will end up with a remainder of 1—that '1' that was added on to give P in the first place. So, if P is not prime, it must be evenly divisible by some prime different from (and hence bigger than) all of $p_1, \ldots, p_n$. In particular, there *must be* a prime bigger than all of $p_1, \ldots, p_n$, so again, the sequence can be continued.

It is interesting to observe that, when you look at the number

$$P = p_1 \times p_2 \times \ldots \times p_n + 1$$

used in Euclid's proof, you don't actually know whether P is itself prime or not. The proof uses two arguments, one that works when P is prime, one that works when it is not. An obvious question to ask is whether it is always one or the other.

The first few values of P look like this:

$P_1 = 2 + 1 = 3$

$P_2 = 2 \times 3 + 1 = 7$

$P_3 = 2 \times 3 \times 5 + 1 = 31$

$P_4 = 2 \times 3 \times 5 \times 7 + 1 = 211$

$P_5 = 2 \times 3 \times 5 \times 7 \times 11 + 1 = 2{,}311$

These are all prime numbers. But the next three values are not prime:

$P_6 = 59 \times 509$

$P_7 = 19 \times 97 \times 277$

$P_8 = 347 \times 27{,}953$

It is not known whether the number P_n is prime for infinitely many values of n. Nor is it known if infinitely many of the numbers P_n are composite. (Of course, at least one of these two alternatives must be true. Most mathematicians would guess that both are in fact true.)

Returning to the prime density function $\pi(N)/N$, one obvious question is whether there is a *pattern* to the way the density decreases as N gets bigger.

There is certainly no simple pattern. No matter how high up through the numbers you go, you keep finding groups of two or more primes clustered closely together, as well as long stretches that are barren of primes altogether. Moreover, these clusters and barren regions seem to occur in a random fashion.

In fact, the distribution of primes is not completely chaotic. But nothing was known for certain until well into the nineteenth century. In 1850, the Russian mathematician Pafnuti Chebychef managed to prove that between any number N and its double $2N$, you can always find at least one prime. So there is *some* order to the way the primes are distributed.

As it turns out, there is considerable order, but you have to look hard to find it. In 1896, the French mathematicians Jacques Hadamard and Charles de la Vallée Poussin independently proved the remarkable result that, as N gets bigger, the prime density $\pi(N)/N$ gets closer and closer to the quantity $1/\ln N$ (where ln is the natural logarithm function, discussed in Chapter 3). This result is nowadays referred to as the *prime number theorem*. It provides a remarkable connection between the natural numbers, which are the fundamental fabric of counting and arithmetic, and the natural logarithm function, which has to do with real numbers and calculus (see Chapter 3).

Over a century before it was proved, the prime number theorem had been suspected by the fourteen-year-old mathematical child prodigy Karl Friedrich Gauss. So great were Gauss's achievements that he deserves an entire section all to himself.

The child genius

Born in Brunswick, Germany, in 1777, Karl Friedrich Gauss (Figure 1.6) displayed immense mathematical talent from a very early age. Stories tell of him being able to maintain his father's business accounts at age three. In elementary school, he confounded his teacher by observing a pattern that enabled him to avoid a decidedly tedious calculation.

Gauss's teacher had asked the class to add together all the numbers from 1 to 100. Presumably the teacher's aim was to keep the students occupied for a time while he was engaged in something else. Unfortunately for the teacher, Gauss quickly spotted the following shortcut to the solution.

You write down the sum twice, once in ascending order, then in descending order, like this:

$$1 + 2 + 3 + \ldots + 98 + 99 + 100$$
$$100 + 99 + 98 + \ldots + 3 + 2 + 1$$

Now you add the two sums, column by column, to give

$$101 + 101 + 101 + \ldots + 101 + 101 + 101$$

There are exactly 100 copies of the number 101 in this sum, so its value is $100 \times 101 = 10{,}100$. Since this product represents twice the answer

FIGURE **1.6** Karl Friedrich Gauss (1777–1855).

to the original sum, if you halve it, you obtain the answer Gauss's teacher was looking for, namely, 5,050.

Gauss's trick works for any number n, not just 100. In the general case, when you write the sum from 1 to n in both ascending and descending order and then add the two sums column by column, you end up with n copies of the number $n + 1$, which is a total of $n(n + 1)$. Halving this total gives the answer:

$$1 + 2 + 3 + \ldots + n = n(n + 1)/2.$$

This formula gives the general pattern of which Gauss's observation was a special case.

It is interesting to note that the formula on the right-hand side of the above identity also captures a geometric pattern. Numbers of the form $n(n + 1)/2$ are called *triangular numbers*, since they are exactly the numbers you can obtain by arranging dots in an equilateral triangle. The first five triangular numbers, 1, 3, 6, 10, 15, are shown in Figure 1.7.

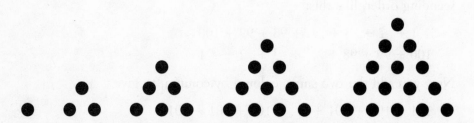

FIGURE **1.7** The numbers 1, 3, 6, 10, 15, . . . are called triangular numbers because they give the numbers of dots that can be arranged in an equilateral triangle.

Gauss's clock arithmetic

In 1801, when Gauss was just twenty-four years old, he wrote a book called *Disquisitiones Arithmeticae*, which to this day is regarded as one of the most influential mathematics books ever written. One of the topics Gauss examined in that book is the idea of finite arithmetic.

You get a finite arithmetic whenever you use a counting system that periodically cycles back on itself and starts again. For instance, when you tell the time, you count the hours 1, 2, 3, and so on, but when you reach 12 you start over again, 1, 2, 3, and so on. Similarly, you count minutes from 1 to 60 and then start over again. This use of finite arithmetic to tell the time is why it is sometimes referred to as 'clock arithmetic'. Mathematicians generally refer to it as modular arithmetic.

To turn the familiar concept of counting minutes and hours into a piece of proper mathematics, Gauss found that he had to change the numbering a little, and start counting from 0. Using Gauss's version, you would count hours as 0, 1, 2, up to 11, and then start over again at 0; minutes would be counted 0, 1, 2, up to 59, and then you would start back at 0.

Having made this minor modification, Gauss investigated the arithmetic of such number systems. The results are often simple, and occasionally quite startling. For example, in the case of hours arithmetic, if you add 2 and 3 you get 5 (three hours after two o'clock is five o'clock), and if you add 7 and 6 you get 1 (six hours after seven o'clock is one o'clock). This is familiar enough. But if you write down the sums using standard arithmetical notation, the second one looks strange:

$$2 + 3 = 5, \qquad 7 + 6 = 1.$$

In the case of minutes arithmetic, 0 minutes after 45 minutes past the hour is 45 minutes past the hour, and 12 minutes after 48 minutes past the hour is 0 minutes past the hour. These two sums look like this:

$$45 + 0 = 45, \qquad 48 + 12 = 0.$$

For all its strangeness, writing out 'clock arithmetic' in this fashion was a smart move on Gauss's part. It turns out that almost all the rules of ordinary arithmetic are true for finite arithmetic, a classic case of a mathematical pattern carrying over from one area to another (in this case, from ordinary arithmetic to finite arithmetic).

In order to avoid confusing addition and multiplication in finite arithmetic with that in ordinary arithmetic, Gauss replaced the equality symbol with ≡ and referred to this relation not as 'equality', but as *congruence*. So, the first two arithmetical results above would be written as

$$2 + 3 \equiv 5, \qquad 7 + 6 \equiv 1.$$

The number at which you start over again, 12 or 60 in the two examples considered, is called the *modulus* of the arithmetic. Obviously, there is nothing special about the number 12 or 60; these are just the values familiar in telling the time. For any natural number n, there is a corresponding finite arithmetic, the *modular arithmetic* of modulus n, where the numbers are 0, 1, 2, . . . , $n - 1$, and where, when you add or multiply numbers, you discard all whole multiples of n.

I did not give any examples of multiplication above, since we never multiply hours or times of the day. But modular multiplication makes perfect sense from a mathematical point of view. As with addition, you just perform the multiplication in the usual way, but then discard all multiples of the modulus n. So, for example, with modulus 7:

$$2 \times 3 \equiv 6, \qquad 3 \times 5 \equiv 1.$$

Gauss's notion of congruence is often used in mathematics, sometimes with several different moduli at the same time. When this is the case, in order to keep track of the modulus being used on each occasion, mathematicians generally write congruences like this:

$$a \equiv b \ (\text{mod } n)$$

where n is the modulus concerned for this particular congruence. This expression is read as "a is congruent to b modulo n."

For any modulus, the operations of addition, subtraction, and multiplication are entirely straightforward. (I did not describe subtraction above, but it should be obvious how it works: in terms of the two clock arithmetics, subtraction corresponds to counting backward in time.) Division is more of a problem: sometimes you can divide, sometimes you cannot.

For example, in modulus 12, you can divide 7 by 5, and the answer is 11:

$$7/5 \equiv 11 \ (\text{mod } 12).$$

To check this, multiply back up by 5, to give

$$7 \equiv 5 \times 11 \ (\text{mod } 12),$$

which is correct, since when you discard multiples of 12 from 55, you are left with 7. But in modulus 12 it is not possible to divide any number by 6, apart from 6 itself. For example, you cannot divide 5 by 6. One way to see this is to observe that if you multiply any of the numbers from 1 to 11 by 6, the result will be an even number, and hence cannot be congruent to 5 modulo 12.

However, in the case where the modulus n is a prime number, division is always possible. So, for a prime modulus, the corresponding modular arithmetic has all the familiar properties of ordinary arithmetic performed with the rational or the real numbers; in mathematician's language, it is a *field*. (Fields appear again on page 73.) And there you have yet another pattern: the pattern that connects the primes with the ability to perform division in modular arithmetic. Which brings us to the topic of patterns of primes, and with it the greatest of all amateur mathematicians: Pierre de Fermat.

The great amateur

Fermat (Figure 1.8), who lived in France from 1601 to 1665, was a lawyer by profession, attached to the provincial parliament at Toulouse. It was not until he was in his thirties that he took up mathematics, as a hobby. It turned out to be quite a hobby: For instance, in addition to making a number of highly significant discoveries in number theory, he developed a form of algebraic geometry some years before René Descartes, who is generally credited with the discovery of how to use algebra to solve problems in geometry. Fermat also founded the subject of probability theory in correspondence with Blaise Pascal, and he laid much of the groundwork for the development of the differential calculus, which was to come to fruition a few years later with the work of Gottfried Leibniz and Isaac Newton. Each of these was a major achievement. Fermat's greatest fame, however, comes from his uncanny ability at finding patterns in the natural numbers, generally patterns involving primes. In fact, he did not just spot these patterns; in most cases he was able to prove conclusively that his observations were correct.

As an amateur mathematician, Fermat published little of his work. What is known of his many accomplishments comes largely from the writings of others, for he made up in letter writing what was not forthcoming by way of publication, maintaining a regular correspondence with some of the finest mathematicians in Europe.

For instance, in a letter he wrote in 1640, Fermat observed that if a is any natural number and p is a prime that does not divide evenly into a, then p has to divide $a^{p-1} - 1$.

For example, take $a = 8$ and $p = 5$. Since 5 does not divide 8, 5 must divide $8^4 - 1$, according to Fermat's observation. If you work out this

FIGURE **1.8** Pierre de Fermat, "The Great Amateur" (1601–1665).

number, you obtain $8^4 = 4,096 - 1 = 4,095$, and indeed, you can see at once that 5 does divide this number. Similarly, it must also be the case that 19 divides $145^{18} - 1$, though in this case most people would be justifiably hesitant to try to check the result by direct calculation.

Though perhaps not apparent at first encounter, Fermat's observation turns out to have several important consequences, not only in mathematics but also in other applications (among them the design of certain data encryption systems and a number of card tricks). In fact, this result crops up so often that mathematicians have given it a name: it is called *Fermat's little theorem*. Nowadays, a number of highly ingenious proofs of this theorem are known, but no one has any idea how Fermat himself proved it. As was his habit, he kept his own methods secret, offering the result as a challenge to others. In the case of his 'little theorem', a complete proof was not found until 1736, when the great Swiss mathematician Leonard Euler finally rose to meet Fermat's challenge.

Fermat's little theorem can be reformulated in terms of modular arithmetic as follows. If p is a prime, and a is any number between 1 and $p - 1$ inclusive, then

$$a^{p-1} \equiv 1 \pmod{p}.$$

Taking the case $a = 2$, for any prime p greater than 2,

$$2^{p-1} \equiv 1 \pmod{p}.$$

Consequently, given a number p, if the above congruence fails, then p cannot be prime. This observation provides an efficient way to try to determine whether a given number is prime, as I will now show.

Taking the prime test

The most obvious method of testing whether a number N is prime is to look for prime factors. To do this, you may have to trial-divide N by all prime numbers up to $\sqrt{N}$. (You do not need to search beyond $\sqrt{N}$, since if N has a prime factor, it must have one no bigger than $\sqrt{N}$.) For fairly small numbers, this is a reasonable approach. Given a moderately powerful computer, the calculations will run virtually instantaneously for any number with 10 or fewer digits. For instance, if N has 10 digits, $\sqrt{N}$ will have 5 digits, and so will be less than 100,000. So, referring to the table on page 27, there are fewer than 10,000 primes to generate and trial-divide into N. This is child's play for a modern computer capable of performing over a billion arithmetic operations a second. But even the most powerful computer could take up to two hours to cope with a 20-digit number, and a 50-digit number could require 10 billion years. Of course, you could strike it lucky and hit a prime divisor fairly soon; the problem occurs when the number N is prime, since you will then have to test prime factors all the way to $\sqrt{N}$ before you are done.

So, testing whether a number is prime by trial division is not feasible for numbers having many more than 20 digits. By looking for patterns in the primes, however, mathematicians have been able to devise a number of alternative means to determine whether a given number is prime. Fermat's little theorem provides one such method. To test whether a given number p is prime using Fermat's little theorem, you compute 2^{p-1} in mod p arithmetic. If the answer is anything other than 1, you know that p cannot be prime. But what happens if the answer does turn out to be 1? Unfortunately, you cannot conclude that p is prime. The problem is that, although $2^{p-1} \equiv 1 \pmod{p}$ whenever p is prime, there are also some *nonprimes* p for which this is true. The smallest such number is 341, which is the product of 11 and 31.

The method would still be fine if 341 were just one of a handful of such numbers, since you could check to see whether p was one of those awkward ones. Unfortunately, there are infinitely many awkward numbers. Thus, Fermat's little theorem is reliable only for showing that a particular number is composite: if the congruence $2^{p-1} \equiv 1 \pmod{p}$ fails, the number p is definitely composite. On the other hand, if the congruence is true, then p may be prime, but then again it may not. If you're feeling lucky, you might want to take a chance and assume that the number is prime anyway, and the odds will be on your side. Composite numbers p for which $2^{p-1} \equiv 1 \pmod{p}$ are fairly rare; there are just two below 1,000—namely, 341 and 561—and only 245 below 1,000,000. But, since there are infinitely many such rare numbers altogether, mathematically speaking, it is not a safe bet that p is prime, and a far cry from mathematical certainty.

Thus, Fermat's little theorem fails to provide a completely reliable means of testing whether a number is prime because of the uncertainty that arises when $2^{p-1} \equiv 1 \pmod{p}$. In 1986, the mathematicians L. M. Adleman, R. S. Rumely, H. Cohen, H. W. Lenstra, and C. Pomerance found a way to eliminate this uncertainty. Starting with Fermat's little theorem, they developed what has turned out to be one of the best general-purpose methods available today for testing whether a number is prime. If you run this test, known as the ARCLP test, on a fast supercomputer, it will take less than 10 seconds for a 20-digit number and less than 15 seconds for a 50-digit number.

The ARCLP test is completely reliable. It is referred to as 'general purpose' because it can be used on any number N. A number of primality tests have been devised that work only on numbers of particular forms, such as numbers of the form $b^n + 1$, for some b, n. In such special cases, it may be possible to handle numbers that, in terms of sheer size, would defeat even the ARCLP test.

Keeping secrets

The ability to find large prime numbers became significant outside mathematics when it was discovered that large primes could be used to encrypt messages sent by insecure channels such as telephone lines or radio transmission. Here, in general outline, is how the method works.

Using a fast computer, together with a primality test such as the ARCLP test, it is an easy matter to find two prime numbers, each about 75 digits in length. The computer can also multiply the two primes together to obtain a composite number of 150 digits. Suppose you were now to give this 150-digit number to a stranger and ask him to find its prime factors. Even if you told him that the number was a product of just two very large primes, it is extremely unlikely that he would be able to complete the assignment, regardless of what computer he had available. For while testing a 150-digit number to see whether it is prime is a task that can be performed in seconds, the best factoring methods known would take impossibly long—years, if not decades or centuries—for numbers of that size, even on the fastest computers available.

Factoring large numbers may be difficult, but not because mathematicians have failed to invent clever methods for the task. Indeed, in recent years some extremely ingenious methods of factoring large numbers have been devised. Using the most powerful computers currently available, it is now possible to factor a number of around 80 digits in a few hours. Since the naive method of trial division could take billions of years for a single 50-digit number, this is already an achievement. But whereas the best available primality tests can cope with numbers having a thousand digits, there are no factoring methods known that can begin to approach this performance level, and indeed, there is some evidence that there may be no such methods: factoring may be an intrinsically more difficult computational task than testing whether a number is prime.

It is this huge disparity between the size of numbers that can be shown to be prime and the size of numbers that can be factored that mathematicians exploit in order to devise one of the most secure forms of 'public key' cipher systems known.

A typical modern cipher system for encrypting messages that have to be sent over insecure electronic communication channels is diagrammed in Figure 1.9. The basic components of the system are two computer programs, an encryptor and a decryptor. Because the design of cipher systems is a highly specialized and time-consuming business, it would be impractical, and probably very insecure, to design separate programs for every customer. Thus, the basic encryption/decryption software tends to be available 'off the shelf', for anyone to purchase. Security for the sender and receiver is achieved by requiring a numerical key for both encryption and decryption. Typically, the key will consist of a number a hundred or more digits long. The security of the system depends on keeping the key secret. For this reason, the users of such systems generally change the key at frequent intervals.

One obvious problem is the distribution of the key. How does one party send the key to the other? To transmit it over the very electronic communication link that the system is supposed to make secure is obviously out of the question. Indeed, the only really safe way is to send the key physically, by trusted courier. This strategy may be acceptable if

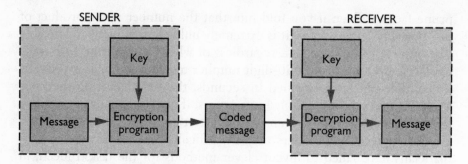

SENDER   RECEIVER

FIGURE 1.9   A typical cipher system.

only two parties are involved, but is completely unfeasible for establishing secure communication between, say, all the world's banks and trading companies. In the financial and business world, it is important that any bank or business be able to contact any other, perhaps at a moment's notice, and be confident that their transaction is secure.

It was in order to meet this requirement that, in 1975, mathematicians Whitfield Diffie and Martin Hellman proposed the idea of a public key cryptosystem (PKS). In a PKS, each potential message receiver (which would be anyone who intends to use the system) uses the software provided to generate not one but *two* keys, an encryption key and a decryption key. The encryption key is published in a publicly available on-line directory. (These days, many people make the public key available on their home page on the World Wide Web.) Anyone who wishes to send a message to a person A looks up A's encryption key, uses that key to encrypt the message, and then sends it off. A then uses her decryption key, which she has divulged to no one, to decrypt the message.

Though the basic idea is simple, actually designing such a system is not. The one originally proposed by Diffie and Hellman turned out to be not as secure as they had thought, but a method devised a short time later by Ronald Rivest, Adi Shamir, and Leonard Adleman has proved to be much more robust. The RSA system, as it is known, is now widely used in the international banking and financial world. (Specially designed RSA computer chips are commercially available.) Here, briefly, is how the RSA system works.

The problem facing the designer of any public key system is this. The encryption process should disguise the message to such an extent that it is impossible to decode it without the decryption key. But since the essence of the system—indeed, of any cipher system—is that the authorized receiver can decrypt the encoded message, the two keys must be mathematically related. Indeed, the receiver's program and decryption key *exactly* undoes the effect of the sender's program and encryption key, so it should be theoretically possible to obtain the decryption key from the encryption key, provided one knows how the cipher program works (and these days, anyone who wants to can find out).

The trick is to ensure that, although it is *theoretically* possible to recover the decryption key from the publicly available encryption key, it is *practically* impossible. In the case of the RSA system, the receiver's secret decryption key consists of a pair of large prime numbers (say, 75 digits each), and the public encryption key consists of their product. Message encryption corresponds (very roughly) to multiplication of the two 75-digit primes; decryption corresponds (equally roughly) to factoring the 150-digit product, a task that is quite unfeasible given present-day knowledge and technology. (The exact encryption and decryption procedures involve a generalization of Fermat's little theorem.)

Easy to guess, hard to prove

Because we are all so familiar with the positive whole numbers—the natural numbers—and because they are so simple, it is easy to find patterns in them. Very often, however, it turns out to be extremely difficult to prove that those patterns are true for all natural numbers. In particular, over the years, mathematicians have proposed numerous simple conjectures about prime numbers that, for all their apparent simplicity, remain unresolved to this day.

One example is the Goldbach conjecture, first raised by Christian Goldbach in a letter to Euler written in 1742. This conjecture proposes that every even number greater than 2 is a sum of two primes. Calculation reveals that this is certainly true for the first few even numbers: $4 = 2 + 2$, $6 = 3 + 3$, $8 = 3 + 5$, $10 = 5 + 5$, $12 = 5 + 7$, and so on. And computer searches have verified the result to at least a billion. But, for all its simplicity, to this day it is not known for certain whether the conjecture is true or false.

Another simple question that no one has been able to answer is the twin primes problem: are there infinitely many pairs of 'twin primes', primes that are just two whole numbers apart, such as 3 and 5, 11 and 13, 17 and 19, or, going a little higher, 1,000,000,000,061 and 1,000,000,000,063?

Still another unsolved problem was first posed by one of Fermat's contemporaries, the French monk Marin Mersenne. In his 1644 book *Cogitata Physica-Mathematica*, Mersenne stated that the numbers

$$M_n = 2^n - 1$$

are prime for $n = 2, 3, 5, 7, 13, 17, 19, 31, 67, 127, 257$, and composite for all other values of n less than 257. No one knows how he arrived at this assertion, but he was not far from the truth. With the arrival of desk calculators, it became possible to check Mersenne's claim, and in 1947 it was discovered that he made just five mistakes: M_{67} and M_{257} are not prime, and M_{61}, M_{89}, M_{107} are prime.

Numbers of the form M_n are nowadays known as Mersenne numbers. Calculation of the first few Mersenne numbers might lead one to suspect that M_n is prime whenever n is prime:

$$M_2 = 2^2 - 1 = 3 \qquad M_3 = 23 - 1 = 7$$
$$M_5 = 2^5 - 1 = 31 \qquad M_7 = 2^7 - 1 = 127$$

all of which are prime. But then the pattern breaks down, with $M_{11} = 2{,}047 = 23 \times 89$. After that, the next Mersenne primes are M_{31}, M_{61}, M_{89}, M_{107}, and M_{127}.

What is the case is the opposite result: that M_n can be prime *only* when n is prime. All it takes to prove this assertion is a bit of elementary algebra. So, in looking for Mersenne primes, it is only necessary to look at Mersenne numbers M_n for which n is itself prime.

Another conjecture that many have been tempted to make on the basis of the numerical evidence is that M_n is prime whenever n is itself a Mersenne prime. The pattern holds until you reach the Mersenne prime $M_{13} = 8{,}191$; the 2,466-digit number $M_{8,191}$ is composite.

The task of finding Mersenne primes is made easier by virtue of a simple, reliable, and computationally efficient method of determining whether a Mersenne number is prime. Called the Lucas–Lehmer test, it is based on Fermat's little theorem. Unlike the ARCLP test, however, the Lucas–Lehmer test works only for Mersenne numbers. On the other hand, whereas the ARCLP test will work only for numbers with a thousand or so digits, the Lucas–Lehmer test has been used to establish the primality of a Mersenne number with almost a million digits: $M_{3,021,377}$.

This gigantic Mersenne number was shown to be prime in 1998 by Roland Clarkson, a 19-year-old mathematics enthusiast in California. Clarkson computed his way into the mathematical record books using a program that he downloaded from the World Wide Web and ran on his home PC. Clarkson's prime, which has exactly 909,526 digits, is the thirty-seventh Mersenne prime to be found.

How big is a number with just short of a million digits? Written out in full, it would fill a 500-page paperback book, it would stretch for more a mile and a half, and it would take a month to say the number, speaking for eight hours a day.

It took Clarkson's PC two weeks to complete the calculation that showed that this number is prime. To make sure the computation was correct, Clarkson asked veteran prime hunter David Slowinski to check the result. A previous holder of a number of record prime discoveries, Slowinski, who works for Cray Research, used a Cray T90 supercomputer to repeat the calculation.

Clarkson is one of over four thousand volunteers who use all their spare computer time to search for Mersenne primes as members of GIMPS, the Great Internet Mersenne Prime Search. GIMPS is a worldwide project coordinated by George Woltman, a programmer living in Orlando, Florida, who wrote and supplies the software. The hunt for record prime numbers used to be the exclusive domain of supercomputers. But by running the software on thousands of individual machines, it is possible to collectively surpass the power of even the world's most powerful supercomputer.

Woltman started GIMPS early in 1996, and it rapidly attracted a substantial number of devotees. Schoolteachers in elementary through high school grades have used GIMPS to get their students excited about doing mathematics. Intel now uses the program to test every Pentium II and Pentium Pro chip before it ships.

Clarkson's discovery was the third success of GIMPS. In November 1996, GIMPS member Joel Armengaud of France found the world record Mersenne prime, $M_{1,398,269}$, and in 1997, Gordon Spence of England topped that with $M_{2,976,221}$.

It is not known whether there are infinitely many Mersenne primes.

Fermat's last theorem

There are many easy-to-state but hard-to-solve problems about numbers besides the ones we have just considered. Undoubtedly, the one that has achieved the greatest fame is Fermat's last theorem. Indeed, Fermat's last theorem and Pythagoras' theorem probably rank as the two best-known theorems of mathematics. And yet it was not until 1994 that Fermat's last theorem was finally proved, after more than three hundred years of effort. Formulated by Fermat, who at one time claimed to have a proof, the theorem was so named when it became the last of Fermat's results for which no proof had been discovered.

The story begins in 1670, five years after Fermat's death. His son Samuel was working his way through his father's notes and letters, with the intention of publishing them. Among these papers, he found a copy of Diophantus' book *Arithmetic*, the 1621 edition, edited by Claude Bachet, who had added a Latin translation alongside the original Greek text. It was this edition that had brought Diophantus' work to the attention of the European mathematicians, and it was clear from the various comments Fermat had written in the margins that the great French amateur had developed his own interest in number theory largely through the study of this master from the third century A.D.

Among Fermat's marginal comments were some forty-eight intriguing, and sometimes significant, observations, and Samuel decided to publish a new edition of *Arithmetic* with all his father's notes as an appendix. The second of these *Observations on Diophantus*, as Samuel called his father's comments, had been written by Fermat in the margin of Book II, alongside Problem 8.

Problem 8 asks: "Given a number which is a square, write it as a sum of two other squares." Written in algebraic notation, this problem says: given any number z, find two other numbers x and y such that

$$z^2 = x^2 + y^2.$$

Fermat's note read as follows:

> On the other hand, it is impossible for a cube to be written as a sum of two cubes or a fourth power to be written as a sum of two fourth powers or, in general, for any number which is a power greater than the second to be written as a sum of two like powers. I have a truly marvelous demonstration of this proposition which this margin is too narrow to contain.

Putting this text into modern notation, what Fermat claimed is that the equation

$$z^n = x^n + y^n$$

has no (whole-number) solutions for any power n greater than 2. (Mathematicians ignore the trivial solutions that arise when one of the unknowns is allowed to be zero.)

So began a saga that continued until late in 1994, as mathematician after mathematician, professional and amateur, attempted to produce a proof—perhaps the same proof Fermat had discovered, if such he had. In fact, it is likely that Fermat was mistaken in his original belief, and subsequently realized his error. His marginal note was not, after all, intended for publication, so he would have had no reason to go back and erase it if he later found a flaw in his reasoning. Certainly, when a proof was finally obtained, it used a significant amount of mathematics not known in Fermat's time (so much so that I will have to delay giving you a sketch of the proof until much later in the book).

But whether or not Fermat had actually found a proof, the story was practically irresistible: a seventeenth-century amateur mathematician claims to have solved a problem that resists three hundred years of attack by the world's finest mathematical minds. Add the fact that the great majority of Fermat's claims did turn out to be correct, plus the simplicity of the statement itself, which any schoolchild can understand, and there is little reason to be surprised at the fame Fermat's last theorem achieved. The offer of a number of large cash prizes for the first person to find a proof only added to its allure: in 1816, the French Academy offered a gold medal and a cash prize, and in 1908 the Royal Academy of Science in Göttingen offered another cash prize, the Wolfskell Prize (worth about $50,000 when it was finally awarded in 1997).

It was its very resistance to proof that led to the theorem's fame. Fermat's last theorem has virtually no consequences, either in mathematics or in the everyday world. In making his marginal note, Fermat was simply observing that a particular numerical pattern that held for square powers did not hold for any higher powers; its interest was purely academic. Had the issue been quickly decided one way or the other, his observation would have been worth nothing more than a footnote in subsequent textbooks.

And yet, had the problem been resolved early on, the mathematical world might have been a great deal the poorer, for the many attempts to solve the problem led to the development of some mathematical notions and techniques whose importance for the rest of mathematics far outweighs that of Fermat's last theorem itself. And when a proof was finally found, by the English-born mathematician Andrew Wiles, it came as a 'simple consequence' of a series of remarkable new results that opened up a completely new area of mathematics.

The Fermat saga begins

As stated earlier, the problem in Diophantus' *Arithmetic* that started the whole affair is that of finding whole-number solutions to the equation

$$z^2 = x^2 + y^2.$$

This problem is obviously related to the Pythagorean theorem. The question can be reformulated in an equivalent geometric form as, do there exist right-angled triangles all of whose sides are a whole number of units in length?

One well-known solution to this problem is the triple $x = 3$, $y = 4$, $z = 5$:

$$3^2 + 4^2 = 5^2.$$

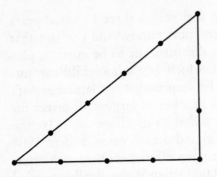

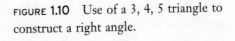

FIGURE 1.10   Use of a 3, 4, 5 triangle to construct a right angle.

This solution was known as far back as the time of ancient Egypt. In fact, it has been claimed that as long ago as 2000 B.C., Egyptian architects used the fact that a triangle whose sides have lengths 3, 4, and 5 units is right-angled in order to construct right angles in buildings. According to this claim, they would first tie twelve equally long pieces of rope into a loop. Then, placing one of the knots at the point where they wanted to construct a right angle, they would pull the loop taut to form a triangle whose sides emanating from the starting point consisted of exactly three and four rope lengths, as shown in Figure 1.10. The resulting triangle would then be right-angled, and so they would have their right angle.

In fact, this trick is not an application of the Pythagorean theorem, but of its converse: if a triangle is such that its sides are related by the equation

$$h^2 = a^2 + b^2,$$

then the angle opposite the side of length h is a right angle. This is Proposition I.48 in *Elements*. The Pythagorean theorem itself is Proposition I.47.

Is 3, 4, 5 the only solution? It does not take long to realize that the answer is no. Once you have found one solution, you immediately get a whole infinite family of solutions, since you can take your first solution and multiply all three values by any number you please, and the result will be another solution. Thus, from the 3, 4, 5 solution, you obtain the solution $x = 6$, $y = 8$, $z = 10$, the solution $x = 9$, $y = 12$, $z = 15$, and so on.

This trivial means of producing new solutions from old can be eliminated by asking for solutions to the equation that have no factor common to all three numbers. Such solutions are generally referred to as *primitive solutions*.

If only primitive solutions are allowed, are there any besides the 3, 4, 5 solution? Again, the answer is well known: the triple $x = 5$, $y = 12$,

$z = 13$ is another primitive solution, and so too is the triple $x = 8$, $y = 15$, $z = 17$.

In fact, there are an infinite number of primitive solutions, and a complete resolution of the problem was given by Euclid in *Elements*, in the form of an exact pattern for all primitive solutions. The formulas

$$x = 2st, \qquad y = s^2 - t^2, \qquad z = s^2 + t^2,$$

generate all the primitive solutions to the original equation, as s and t vary over all natural numbers such that:

1. $s > t$
2. s and t have no common factor
3. one of s, t is even, the other odd.

Moreover, any primitive solution to the equation is of the above form for some values of s, t.

Returning now to Fermat's last theorem itself, there is some evidence to suggest that Fermat did have a valid proof in the case $n = 4$. That is to say, it is possible that he was able to prove that the equation

$$z^4 = x^4 + y^4$$

has no whole-number solutions. The evidence in question consists of one of the few complete proofs Fermat did leave behind: an ingenious argument to demonstrate that the area of a right-angled triangle whose sides are all whole numbers cannot be a square of a whole number. From this result, Fermat was able to deduce that the equation

$$z^4 = x^4 + y^4$$

can have no whole-number solutions, and it is reasonable to assume that he established the result about triangles with square areas precisely in order to deduce this case of his 'last theorem'.

In order to establish his result about areas of triangles, Fermat's idea was to show that if there were natural numbers x, y, z such that

$$z^2 = x^2 + y^2,$$

and if, in addition, $\frac{1}{2}xy = u^2$ for some natural number u (i.e., if the area of the triangle is a square), then there are another four numbers x_1, y_1, z_1, u_1 that stand in the same relation to each other, for which $z_1 < z$.

Then, one may apply the same argument again to produce four more numbers, x_2, y_2, z_2, u_2 that also stand in the same relation and for which $z_2 < z_1$.

But this process can go on forever. In particular, you will end up with an infinite sequence of natural numbers z, z_1, z_2, z_3, . . . , such that

$$z > z_1 > z_2 > z_3 > \ldots$$

But such an infinite sequence is impossible; eventually the sequence must descend to 1, and then it will stop. Hence there can be no numbers x, y, z with the supposed properties, which is what Fermat set out to prove.

For obvious reasons, this method of proof is referred to as Fermat's method of infinite descent. It is closely related to the present-day method of mathematical induction, a powerful tool for verifying many patterns involving natural numbers which I will outline in the next section.

With the case of exponent $n = 4$ established, mathematicians readily observed that if Fermat's last theorem is also true for all *prime* exponents, then it must be true for all exponents. So, the would-be last theorem prover was faced with the case of an arbitrary prime exponent.

The first person to make any real progress in this direction was Euler. In 1753, he claimed to have proved the result for $n = 3$. Though his published proof contained a fundamental flaw, the result is generally still credited to him. The problem with Euler's proof was that it depended upon a particular assumption about factorization that he made in the course of his argument. Though this assumption can in fact be proved for the case $n = 3$, it is not true for all prime exponents, as Euler seemed to be assuming, and in fact it was precisely this subtle, but invalid, assumption that brought down many subsequent attempts to prove Fermat's last theorem.

In 1825, extending Euler's argument, Peter Gustav Lejeune Dirichlet and Adrien-Marie Legendre proved Fermat's last theorem for exponent $n = 5$. (Their version of the argument avoided the factorization trap that befell Euler.)

Then, in 1839, using the same general approach, Gabriel Lamé proved the result for $n = 7$. By this stage, the argument was becoming increasingly intricate, and there seemed little hope of taking it any further to deal with the next case, $n = 11$. (Not that this kind of piecemeal approach would solve the entire problem in any case.)

To make any further progress, what was required was the detection of some kind of general pattern in the proofs, a way of stepping back from the complexity of the trees to the larger-scale order of the forest. This advance was made by the German mathematician Ernst Kummer in 1847.

Kummer recognized that some prime numbers exhibited a certain kind of pattern, which he referred to as *regularity*, that enabled an Euler-type proof of Fermat's last theorem to be carried through. Using this new property of regularity, Kummer was able to prove that Fermat's last theorem holds for all exponents n that are regular primes. Of the primes less than 100, only 37, 59, and 67 fail to be regular, so in one fell swoop Kummer's result established Fermat's last theorem for all exponents up to 36 and for all prime exponents less than 100 except for 37, 59, and 67.

There are a number of different, though totally equivalent, ways of defining exactly what a regular prime is, but all refer to some fairly advanced mathematical concepts, so I will not give any definition here. I will simply tell you that, by the end of the 1980s, computer searches as far as 4,000,000 had shown that most primes are regular. Moreover, all the nonregular primes less than 4,000,000 were shown to satisfy a property a bit weaker than regularity, but which still implied Fermat's last theorem for these exponents. So by the start of the 1990s, Fermat's last theorem was known to be true for all exponents up to 4,000,000.

At this point we must leave Fermat's last theorem, but only for the time being. We shall come back to it in Chapter 6, when I shall tell you about a startling discovery made in 1983, undoubtedly the most significant advance on Fermat's last theorem subsequent to Kummer's work. I shall also describe a series of unexpected and highly dramatic events that took place from 1986 to 1994, which culminated in the final resolution of the three-hundred-year saga of Fermat's last theorem. The reason for putting off these two developments until later—indeed, several chapters later—is itself a striking illustration that mathematics is the search for, and study of, patterns. Both the 1983 discovery and the events of 1986 to 1994 came about only as a result of investigations of patterns of quite different natures—not number patterns, but patterns of shape and position, patterns that involve the infinite in a fundamental way.

The domino effect

To end this chapter, let me fulfill my earlier promise and tell you a little bit about the method of mathematical induction. (You will recall that when we looked at Fermat's method of infinite descent (page 46), I mentioned that Fermat's clever trick was related to induction.)

The method of mathematical induction is one of the most powerful weapons in the mathematician's arsenal. It enables you to conclude that a pattern holds for *all* natural numbers based on just two pieces of evidence. Just think of the productivity involved: you prove just *two* results and you are able to draw a conclusion about the *infinitely many* natural numbers.

It is easy to appreciate the method in an intuitive way as a 'domino argument'. Suppose you were to stand a row of dominoes on their ends in a row. Let $P(n)$ abbreviate the statement that "the nth domino falls down." Now let me try to convince you that, under certain circumstances, the entire row of dominoes will fall down: that is, $P(n)$ will be true for all n.

First, I tell you that the dominoes are placed sufficiently close together that if one falls, it will knock over the next one. In terms of our abbreviation, I tell you that if $P(n)$ is true for any n, then $P(n + 1)$ will be true. That's the first item of information.

Next, I tell you that the first domino is knocked over. In terms of our abbreviation, I tell you that $P(1)$ is true. That's the second item of information.

On the basis of those two pieces of information, you can clearly conclude that the entire row of dominoes will fall down. That is, you can conclude that $P(n)$ holds for every n (because each domino is knocked over by its predecessor, and in turn knocks down the next one in the row; see Figure 1.11).

Of course, in real life, the row of dominoes will be finite. But exactly the same idea will work in a more abstract setting, where the abbreviation $P(n)$ refers to some other event, one that makes sense for every natural number n, not just a domino falling down. Here is the general idea.

Suppose you have noticed some pattern—call it P—that seems to hold for every natural number n. For example, perhaps when adding together more and more odd numbers, you notice that the sum of the first n odd numbers always seems to work out to be n^2:

$$1 + 3 = 4 = 2^2$$
$$1 + 3 + 5 = 9 = 3^2$$
$$1 + 3 + 5 + 7 = 16 = 4^2$$
$$1 + 3 + 5 + 7 + 9 = 25 = 5^2$$
$$1 + 3 + 5 + 7 + 9 + 11 = 36 = 6^2$$

and so on.

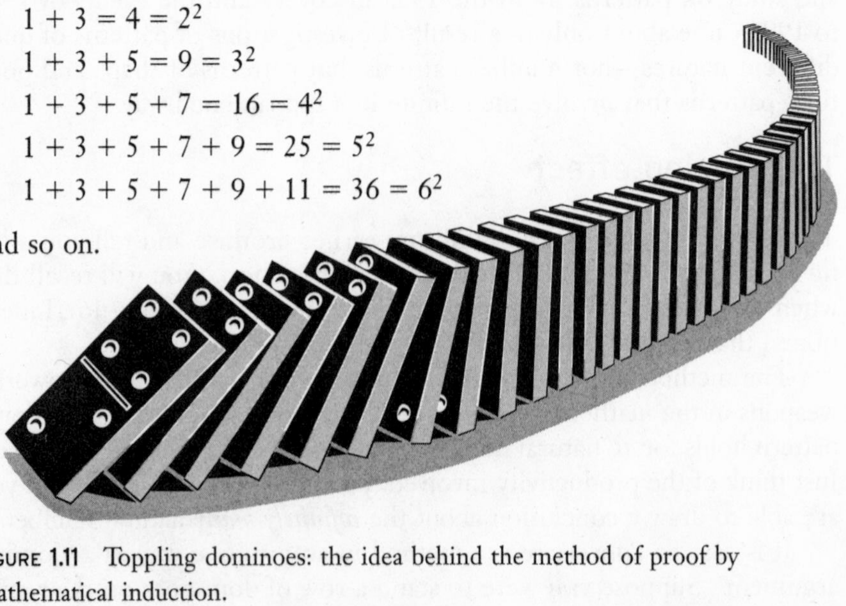

FIGURE 1.11 Toppling dominoes: the idea behind the method of proof by mathematical induction.

You suspect that this pattern continues forever; that is to say, you suspect that for every natural number n, the following identity is true:

$$1 + 3 + 5 + \ldots + (2n - 1) = n^2.$$

Call this particular identity $P(n)$.

How do you prove that the identity $P(n)$ is true for every natural number n? The numerical evidence you have collected may seem pretty convincing—perhaps you have used a computer to verify that $P(n)$ holds for all n up to a billion. But numerical evidence alone can never provide you with a rigorous proof, and indeed, there have been a number of instances in which numerical evidence of a billion or more cases has turned out *not* to be reliable. The problem is, the pattern P you are trying to verify is a pattern over the entire infinitude of all natural numbers. How can you show that a pattern holds for infinitely many objects? Certainly not by checking every single case.

This is where the method of mathematical induction comes in. Just as with the row of dominoes, in order to prove that a property $P(n)$ holds for every natural number n, it is enough to prove just two facts: first, that $P(n)$ holds for the number $n = 1$, i.e., $P(1)$ is true; second, that if $P(n)$ is assumed to be true for an arbitrary number n, then $P(n + 1)$ follows. If you can establish both these facts, then you may conclude, without any further effort, that $P(n)$ is true for every natural number n.

Using the method of induction, it is straightforward to verify the particular pattern P given above. For $n = 1$, the identity just says

$$1 = 1^2,$$

which is trivial. Now assume that the identity holds for an arbitrary number n. (This is like supposing that the nth domino is knocked down.) That is, assume that

$$1 + 3 + 5 + \ldots + (2n - 1) = n^2.$$

The next odd number after $(2n - 1)$ is $(2n + 1)$. Adding this number to both sides of the identity above gives

$$1 + 3 + 5 + \ldots + (2n - 1) + (2n + 1) = n^2 + (2n + 1).$$

By elementary algebra, the expression to the right of the equals sign reduces to $(n + 1)^2$. Hence you can rewrite the last identity as

$$1 + 3 + 5 + \ldots + (2n - 1) + (2n + 1) = (n + 1)^2.$$

This is just the identity $P(n + 1)$. So, the above algebraic argument shows that *if $P(n)$ holds for some n, then $P(n + 1)$ follows*. (In other words, the little algebraic argument I have just given is analogous to my showing you that the nth domino is sufficiently close to the $(n + 1)$st domino to knock it over.)

Thus, by mathematical induction, you may conclude that $P(n)$ does indeed hold for every natural number n.

That's real power. There is no way you could ever verify every single instance of the equation

$$1 + 3 + 5 + \ldots + (2n - 1) = n^2.$$

There are infinitely many such equations. And yet, on the basis of the above proof, you know, absolutely, that *every* such equation is true.

Thus, the method of induction provides mathematicians with a method for establishing that a particular pattern holds for all natural numbers. And this gets us to the very heart of (pure) mathematics ('pure' as opposed to applied mathematics). The pure mathematician is interested in establishing patterns. When a mathematician has observed that some property holds for the first ten natural numbers, or the first hundred, or the first thousand, her natural question is, does this property hold for all natural numbers? Are the observed results indicative of a general pattern? No amount of computational checking can answer this question—it does not help to check, say, the first million cases. All that does is show that the pattern is true for the first million natural numbers. Such evidence might increase our suspicion that there is a general pattern, but it does not constitute a proof. Maybe the pattern breaks down at number 1,000,001. There have been patterns that have been checked by computers for many millions of cases, and yet eventually it has been proved that the pattern breaks down somewhere beyond the cases that have been checked. Of course, as you might suspect, for some applications, knowing that a particular property holds for the first 5 million natural numbers is all you need. If so, then fine—it is enough to use a computer to check those first 5 million cases. But to the pure mathematician, that's not really mathematics, it's merely computation. Mathematics is about the search for 'perfect' patterns. To establish a pattern, the mathematician must find a proof. The method of induction is one of a number of methods that can provide such a proof.

CHAPTER 2

Patterns of the Mind

Proof beyond doubt

From the time of the Greek mathematician Thales onward, proof has played a central role in mathematics. It is by means of proofs that mathematicians determine which statements of mathematics are true (such as Pythagoras' theorem) and which are false. But what exactly is a proof? For example, on page 24, there is a 'proof' of the fact that the number $\sqrt{2}$ is not rational—that is to say, cannot be expressed as a ratio of two whole numbers. Anyone who follows that discussion carefully, and thinks about each step, will surely find it entirely convincing—it does, indeed, *prove* the assertion that $\sqrt{2}$ is irrational. But just what is it about that particular discussion, about those particular English sentences written in that particular order, that makes it a proof?

Admittedly, the argument does make use of some simple algebraic notation, but that is not the crux of the matter. It would be extremely easy to eliminate all the algebra, replacing every symbol with words and phrases of the English language, and the result would still be a proof of the assertion—indeed, it would be *the same proof*! The choice of language, whether symbolic, verbal, or even pictorial, might affect the length of the proof, or the ease with which you can understand it, but it does not affect whether the argument does or does not constitute a proof. In human terms, being a proof means having the capacity to completely convince any sufficiently educated, intelligent, rational person, and surely that capacity has to do with some kind of abstract pattern or abstract structure associated with the argument. What is that abstract structure, and what can be said about it?

Even more fundamental, the same question can be posed about language itself. Proofs are just one of many things that can be expressed by language. What is it about the symbols on this page that enables me, as author, to communicate my thoughts to you, as reader? As in the case of proofs, *the same thoughts* will be communicated by any foreign-language edition of this book, so again the answer is surely not something physical and concrete about this particular page or the particular language used, but has to be some kind of abstract structure associated with what appears on the page. What is this abstract structure?

The abstract patterns that human beings are equipped to recognize and utilize are not to be found solely in the physical world; abstract patterns are also involved in our thinking and communicating with one another.

The logical patterns of Aristotle

The first systematic attempt to describe the patterns involved in proof was made by the ancient Greeks, in particular by Aristotle. These efforts resulted in the creation of what is now called Aristotelian logic. (It is not clear exactly how much of this work is due to Aristotle himself and how much to his followers. Whenever I use the name 'Aristotle' here, I am referring to Aristotle and his followers.)

According to Aristotle, a proof, or rational argument, or logical argument, consists of a series of assertions, each one following logically from previous ones in the series, according to some logical rules. Of course, this description can't be quite right, since it doesn't provide any means for the proof to begin: the first assertion in an argument cannot follow from any previous assertions, since in its case there are no previous assertions! But any proof must depend on some initial facts or assumptions, so the series can start off by listing those initial assumptions, or at least some of them. (In practice, the initial assumptions may be obvious or understood, and might, therefore, not be mentioned explicitly. Here I shall follow normal mathematical practice and concentrate on the ideal case, in which all steps are present.)

Aristotle's next step was to describe the logical rules that may be used to arrive at a valid conclusion. To handle this issue, he assumed that any correct argument may be formulated as a series of assertions of a particular form: the so-called subject–predicate proposition.

A *proposition* is simply a sentence that is either true or false. The subject–predicate propositions considered by Aristotle are those consisting of two entities, a subject and a property, or predicate, ascribed to that subject. Examples of such propositions are:

Aristotle is a man.

All men are mortal.

Some musicians like mathematics.

No pigs can fly.

You might well wonder whether Aristotle was correct in assuming that any valid argument could be broken down into a series of assertions of this particularly simple form. The answer is no; for instance, many mathematical proofs cannot be analyzed in this manner. And even when such an analysis is possible, it can be extremely difficult to actually break the argument down into steps of this kind. So Aristotle's analysis did not, in fact, identify an abstract structure applicable to all correct arguments; rather, his analysis applies only to a certain, very restricted, kind of correct argument.

What gives Aristotle's work lasting historical value nevertheless is that not only did he look for patterns in correct arguments, he actually did find some. It was to be almost two thousand years before anyone took the study of patterns of rational argument significantly further!

The logical rules Aristotle identified that must be followed in order to construct a correct proof (using subject–predicate propositions) are known as *syllogisms*. These are rules for deducing one assertion from exactly two others. An example of a syllogism is:

All men are mortal.

Socrates is a man.

Socrates is mortal.

The idea is that the third assertion, the one below the line, follows logically from the previous two. In the case of this simple—and very hackneyed—example, this deduction certainly seems correct enough, albeit pretty obvious. What makes Aristotle's contribution so significant is that he abstracted a general pattern from such examples.

His first step was to abstract away from any particular example to obtain a general case. Here is the general idea: Let S denote the subject of any subject–predicate proposition, P the predicate. In the case of the proposition *Socrates is a man*, S denotes Socrates and P denotes the predicate 'is a man'. This step is very much like the step of replacing numbers with algebraic symbols such as the letters x, y, and z. But, instead of the symbols S and P denoting arbitrary *numbers*, they denote an arbitrary *subject* and an arbitrary *predicate*. Eliminating the particular in this way sets the stage for examining the abstract patterns of reasoning.

According to Aristotle, the predicate may be used affirmatively or negatively in the proposition, like this:

S is P   or   *S is not P.*

Moreover, the subject may be *quantified*, by expressing it in the form

all S   or   *some S.*

The two kinds of quantification of the subject may be combined with the two possibilities of affirmative and negative predicates to give a total of four possible quantified subject–predicate propositions:

All S is P.
All S is not P.
Some S is P.
Some S is not P.

The second of these is easier to read—and more grammatical—if it is rewritten in the equivalent form

No S is P.

It might also seem more grammatical to write the first of these four propositions in the plural, as

All S are P.

But this is a minor issue that disappears with the next step in the abstraction process, which is to write the four syllogism patterns in abbreviated form:

SaP : All S is P.
SeP : No S is P.
SiP : Some S is P.
SoP : Some S is not P.

These abbreviated forms make very clear the abstract patterns of propositions that Aristotle was looking at.

Most work with syllogisms concentrates on propositions of the four quantified forms just stated. On the face of it, these forms would appear to ignore examples like the one given earlier, *Socrates is a man.* But examples such as this, in which the subject is a single individual, are indeed still covered. In fact, they are covered twice. If *S* denotes the collection of all 'Socrateses', and *P* denotes the property of being a man, then either of the forms *SaP* and *SiP* captures this particular proposition. The point is that, because there is only one Socrates, all of the following are equivalent:

'Socrates',   'all Socrateses',   'some Socrates'.

In everyday English, only the first of these seems sensible. But the whole purpose of this process of abstraction is to get away from ordinary language and work with the abstract patterns expressed in that language.

The decision to ignore individual subjects, and concentrate instead on collections or kinds of subjects, has a further consequence: the subject and predicate in a subject–predicate proposition may be interchanged. For example, *All men are mortal* may be changed to *All mortals are men*. Of course, the changed version will generally mean something quite different from the original, and may be false or even nonsensical; but it still has the same abstract structure as the original, namely, *All 'somethings' are 'somethings'*. In regard to allowable subject–predicate propositions, the four construction rules listed a moment ago are *commutative*: the terms *S* and *P* may be interchanged in each one.

Having described the abstract structure of the propositions that may be used in an Aristotelian argument, the next step is to analyze the syllogisms that may be constructed using those propositions. What are the valid rules that may be used in order to construct a correct argument as a series of syllogisms?

A syllogism consists of two initial propositions, called the *premises* of the syllogism, and a *conclusion* that, according to the rule, follows from the two premises. If *S* and *P* are used to denote the subject and predicate of the conclusion, then, in order for inference to take place, there must be some third entity involved in the two premises. This additional entity is called the middle term; I'll denote it by *M*.

For the example

All men are mortal.
<u>*Socrates is a man.*</u>
Socrates is mortal.

S denotes Socrates, *P* denotes the predicate of being mortal, and *M* is the predicate of being a man. In symbols, this particular syllogism thus has the form

MaP
<u>*SaM*</u>
SaP

(The second premise and the conclusion could also be written using i instead of a.) The premise involving M and P is called the *major premise*, and is written first; the other one, involving S and M, is called the *minor premise*, and is written second.

Having standardized the way syllogisms are represented, an obvious question to ask is, how many possible syllogisms are there?

Each major premise can be written in one of two orders, with the M first or the P first. Similarly, there are two possible orders for the minor premise, S first or M first. The possible syllogisms thus fall into four distinct classes, known as the four *figures* of the syllogism:

I	II	III	IV
MP	PM	MP	PM
SM	SM	MS	MS
SP	SP	SP	SP

For each figure, the gap between the subject and the predicate of each proposition can be filled with any one of the four letters a, e, i, or o. So that gives a total of $4 \times 4 \times 4 \times 4 = 256$ possible syllogisms.

Of course, not all possible patterns will be logically valid, and one of Aristotle's major accomplishments was to find all the valid ones. Of the 256 possible syllogism patterns, Aristotle's list of the valid ones consisted of precisely the nineteen listed below. (Aristotle made two errors, however. His list contains two entries that do not correspond to valid inferences, as the next section will show.)

Figure I: *aaa, eae, aii, eio*
Figure II: *eae, aee, eio, aoo*
Figure III: *aai, iai, aii, eao, oao, eio*
Figure IV: *aai, aee, iai, eao, eio*

How Euler circled the syllogism

Euler invented an elegant way to check the validity of syllogisms using simple geometric ideas. It is called the method of Euler circles. The idea is to represent the syllogism by three overlapping circles, as shown in Figure 2.1. The region inside the circle marked S represents all the objects of type S, and analogously for the circles marked P and M. The procedure used to verify the syllogism is to see what the two premises say about the various regions in the diagram, numbered 1 to 7.

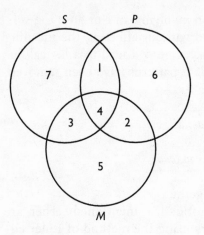

FIGURE 2.1 Euler circles.

To illustrate the method, consider the simple example we considered earlier, namely, the syllogism

MaP
SaM
SaP

The major premise, *All M are P*, says that regions 3 and 5 are empty. (All the objects in *M* are in *P*—that is to say, are in the regions marked 2 and 4.) The minor premise, *All S are M*, says that regions 1 and 7 are empty. Thus, the combined effect of the two premises is to say that regions 1, 3, 5, and 7 are empty.

The goal now is to construct a proposition involving *S* and *P* that is consistent with this information about the various regions. Regions 3 and 7 are empty, so anything in *S* must be in region 1 or region 4, and is therefore in *P*. In other words, all *S* are *P*. And that verifies this particular syllogism.

Though not all syllogisms are as easy to analyze as this example, all the other valid syllogisms can be verified in the same way.

For all its simplicity, the method of Euler circles is remarkable in that it provides a geometric way of thinking about deduction. In the above discussion, patterns of thought were transformed first into algebraic patterns and then into simple geometric patterns. This demonstrates yet again the incredible power of the mathematical method of abstraction.

Now that we have narrowed down the collection of possible syllogisms to the valid ones, we can further simplify Aristotle's list by removing any syllogism whose logical pattern is duplicated elsewhere in the list. For instance, in any proposition involving an *e* or an *i*, the subject and predicate may be interchanged without affecting the meaning of the proposition, and each such interchange will lead to a logical duplicate. When all of the various redundant patterns have been removed, only the following eight forms are left:

Figure I: *aaa, eae, aii, eio*
Figure II: *aoo*
Figure III: *aai, eao, oao*

The fourth figure has disappeared altogether.

The two invalid syllogisms in Aristotle's list, mentioned earlier, are still there. Now can you spot them? Try using the method of Euler circles to check each one. If you fail to find the mistakes, take heart: it took almost two thousand years for the errors to come to light. The next section explains how the mistake was corrected. The length of time it took to resolve the issue was not due to any lack of attention to syllogisms. Indeed, over the centuries, Aristotle's logical theory achieved an exalted status in human learning. For instance, as recently as the fourteenth century, the statutes of the University of Oxford included the rule "Bachelors and Masters of Arts who do not follow Aristotle's philosophy are subject to a fine of five shillings for each point of divergence." So there!

An algebra of thought

From the time of the ancient Greeks until the nineteenth century, virtually no advances were made in the mathematical study of the patterns of rational argument. The first major breakthrough since Aristotle came with the arrival on the mathematical scene of the Englishman George Boole (see Figure 2.2), who found a way to apply the methods of algebra to the study of human reasoning.

Born in 1815 to Irish parents in East Anglia, Boole grew to mathematical maturity at a time when mathematicians were starting to realize that algebraic symbols could be used to denote entities other than numbers, and that the methods of algebra could be applied to domains other than ordinary arithmetic. For instance, the end of the eighteenth century had seen the development of the arithmetic of complex numbers (a generalization of the real numbers, which we will encounter in Chapter 3), as well as Hermann Grassmann's development of vector algebra. (A vector is an entity that has both magnitude and direction, such as velocity or force. Vectors can be studied both geometrically and algebraically.)

FIGURE **2.2** George Boole (1815–1864).

Boole set about trying to capture the patterns of thought in an algebraic fashion. In particular, he sought an algebraic treatment of Aristotle's syllogistic logic. Of course, it is an easy matter simply to represent Aristotle's own analysis using algebraic symbols, as I did in the previous section. But Boole went much further: he provided an algebraic treatment of the logic, using not only algebraic notation but algebraic structure as well. In particular, he showed how to write down equations in his algebra, and how to solve them—and what those equations and their solutions meant in terms of logic.

Boole's brilliant analysis was described in a book he published in 1854, titled *An Investigation of the Laws of Thought on Which are Founded the Mathematical Theories of Logic and Probabilities*. Often referred to more simply as *The Laws of Thought*, this pivotal book built upon an earlier, and much less well known, treatise of Boole's, titled *The Mathematical Analysis of Logic*.

Chapter 1 of *The Laws of Thought* begins with these words:

> The design of the following treatise is to investigate the fundamental laws of those operations of the mind by which reasoning is performed; to give expression to them in the symbolic language of a Calculus, and upon this foundation to establish the science of Logic and construct its method.

The starting point of Boole's logic is the same idea that led to the method of Euler circles; namely, to regard propositions as dealing with classes or collections of objects, and to reason with those collections. For example, the proposition *All men are mortal* can be taken to mean that the class of all men is a subclass (or subcollection, or part) of the class of all mortals. Or, to put it another way, the members of the class of all men are all members of the class of all mortals. But Boole did not look for structure at the level of the members of these classes; rather, he concentrated on the classes themselves, developing an 'arithmetic' *of* classes. His idea is both simple and elegant, and was to prove extremely powerful.

Start off by using letters to denote arbitrary collections of objects, say, x, y, z. Denote by xy the collection of objects common to both x and y, and write $x + y$ to denote the collection of objects that are in either x or y or in both. (Actually, in defining the 'addition' operation, Boole distinguished the case in which x and y have no members in common from the case in which they overlap. Modern treatments—and this book is one such—generally do not make this distinction.)

Let 0 denote an empty collection, and let 1 denote the collection of all objects. Thus, the equation $x = 0$ means that x has no members. The collection of all objects not in x is denoted by $1 - x$.

Boole observed that his new 'arithmetic' of collections has the following properties:

$$x + y = y + x \qquad\qquad xy = yx$$
$$x + (y + z) = (x + y) + z \qquad x(yz) = (xy)z$$
$$x(y + z) = xy + xz$$
$$x + 0 = x \qquad\qquad 1x = x$$
$$2x = x + x = x \qquad\qquad x^2 = xx = x$$

The first five of these identities are familiar properties of ordinary arithmetic, in which the letters denote numbers; they are the two commutative laws, the two associative laws, and the distributive law.

The next two identities indicate that 0 is an identity operation for 'addition' (that is, Boole's 0 behaves like the number 0) and 1 is an identity operation for 'multiplication' (that is, Boole's 1 behaves like the number 1).

The last two identities appear quite strange when you see them for the first time; they certainly are not true for ordinary arithmetic. They are called the idempotent laws.

In modern terminology, any collection of objects and any two operations ('multiplication' and 'addition') on them that obey all of the above identities is called a Boolean algebra. In fact, the system just described is not exactly the one Boole himself developed; in particular, his treatment of 'addition' meant that his system did not include the idempotent law for addition. As is often the case in mathematics—and, indeed, in any walk of life—there is always room for others to improve upon a good idea; in this case, Boole's system was subsequently modified to give the one described here.

Boole's algebraic logic provides an elegant way to study Aristotle's syllogisms. In Boole's system, the four kinds of subject–predicate propositions considered by Aristotle can be expressed like this:

$$SaP : s(1 - p) = 0$$
$$SeP : sp = 0$$
$$SiP : sp \neq 0$$
$$SoP : s(1 - p) \neq 0$$

With the syllogisms represented this way, it is a matter of simple algebra to determine which are valid. For example, take a syllogism in which the two premises are:

All P are M.

No M is S.

Expressing these algebraically, you get the two equations

$$p(1 - m) = 0,$$
$$ms = 0.$$

By ordinary algebra, the first of these can be rewritten like this:

$$p = pm.$$

Then, by playing around for a few moments to find an expression involving just p and s (the conclusion must not involve the middle term), you discover that

$$ps = (pm)s = p(ms) = p0 = 0.$$

In words:

No P is S.

And that is really all there is to using Boole's logic. Everything has been reduced to elementary algebra, except that the symbols stand for propositions rather than numbers (just as Grassmann was able to reduce arguments about vectors to algebra).

It was by using Boole's algebraic logic that two mistakes were found in Aristotle's original treatment of the syllogism. Two of the forms listed by Aristotle as valid are not, in fact, valid. They are both in the third figure: *aai* and *eao*.

In words, the first of these forms says:

All M are P.
All M are S.
Some S is P.

Written algebraically, it looks like this:

$$m(1 - p) = 0$$
$$m(1 - s) = 0$$
$$sp \neq 0$$

The question then becomes, does the third equation follow from the first two? The answer is no. If $m = 0$, then the first two equations are true, *whatever s and p* denote. Consequently, it is possible for the two premises to be true when the conclusion is false, so the form is invalid.

The same kind of thing happens with the other syllogism that Aristotle misclassified.

It is true that, if $m \neq 0$, both syllogisms work out fine. And this is undoubtedly why the error was not spotted for well over a thousand years. When you are thinking in terms of words, about *predicates*, it is not natural to ask what happens if one of the predicates describes something that is impossible. But when you are manipulating simple algebraic equations, it is not only natural but, for the mathematician, second nature to check whether terms are zero or not.

The point is, translating patterns of logic into patterns in algebra does not change those patterns in an intrinsic way. But it does change the way people can think about those patterns. What is unnatural and difficult in one framework can become natural and easy in another. In mathematics, as in other walks of life, it is often not just *what* you say that matters, but the *way* that you say it.

The atomic approach to logic

Boole's algebraic system succeeded, and succeeded extremely well, in capturing Aristotle's syllogistic logic. But its importance rests on far more than that one achievement.

For all its intrinsic interest, Aristotle's system is too narrow in its scope. Though many arguments can be recast as a series of subject–predicate propositions, this is not always the most natural way to express an argument, and besides, many arguments simply cannot be made to fit the syllogistic mold.

Starting with Boole's audacious step of applying algebra to the patterns of reasoning, logicians took a much more general approach to finding the patterns used in deduction. Instead of studying arguments involving propositions of a particular kind, as Aristotle had done, they allowed any proposition whatsoever. In so doing, they resurrected an approach to logic that had been developed by a rival Greek school to Aristotle's, the Stoics, but which had been largely forgotten.

In the Stoic approach, you start off with some basic, unanalyzed propositions. The only thing you know about these propositions is that they *are* propositions; that is, they are statements that are either true or false (though, in general, you don't know which). A number of precisely stipulated rules (outlined below) enable you to combine these basic propositions to produce more complex propositions. You then analyze arguments that consist of a series of such compound propositions.

This system is known nowadays as *propositional logic*. It is highly abstract, since the logical patterns uncovered are entirely devoid of any context. The theory is completely independent of what the various propositions say. It is very much like the molecular theory of matter, in which you regard molecules as being made up of different atoms, but you leave those atoms unanalyzed and concentrate solely on the way they fit together.

Like Aristotle's syllogistic logic, propositional logic still has the defect of being too restrictive; not all arguments are of this kind. Nevertheless, a great many arguments can be analyzed with propositional logic. Moreover, the logical patterns that are uncovered by this approach provide considerable insight into the notion of mathematical proof and, indeed, of logical deduction in general. Since the theory is independent of what the various propositions are about, the patterns uncovered are those of *pure* logic.

Most of the rules used today for combining propositions to give more complex compound propositions are essentially those considered by the Stoics (and later by Boole). The description I shall give here, however, takes into account subsequent refinements that have been made over the years.

Since the only fact known about a proposition is that it is true or false, it is hardly surprising that the notion of truth and falsity plays a central role in this theory. The logical patterns that arise when propositions are combined are patterns of truth.

For instance, one way to combine propositions is by the operation of *conjunction*: given propositions p and q, form the new proposition [p and q]. For example, the conjunction of the two propositions *John likes ice cream* and *Mary likes pineapple* is the compound proposition *John likes ice cream and Mary likes pineapple*. In general, all you can hope to know about the compound proposition [*p and q*] is what its truth status is, given the truth status of *p* and *q*. A few moments reflection gives the appropriate pattern. If both *p* and *q* are true, then the conjunction [*p and q*] will be true; if one or both of *p* and *q* are false, then [*p and q*] will be false.

The pattern is perhaps most clearly presented in tabular form, as what we call a *truth table*. Here are the truth tables for conjunction and for three other operations on propositions, namely, *disjunction* [*p or q*], the *conditional* [*p → q*], and *negation* [*not p*].

p	*q*	*p and q*		*p*	*q*	*p or q*
T	T	T		T	T	T
T	F	F		T	F	T
F	T	F		F	T	T
F	F	F		F	F	F

p	*q*	*p → q*		*p*	*not p*
T	T	T		T	F
T	F	F		F	T
F	T	T			
F	F	T			

In the tables, T denotes the truth value 'true', and F denotes the value 'false'. Reading along a row, each table indicates the truth value of the compound proposition that arises from the truth values of the components. The tables provide the formal definitions of these logical operations.

The last of these, [*not p*], is self-explanatory, but the other two require some comment. In everyday language, the word 'or' has two meanings. It can be used in an exclusive way, as in the sentence "The door is locked *or* it is not locked." In this case, only one of these possibilities can be true. Alternatively, it can be used inclusively, as in "It will rain *or* snow." In this case, there is the possibility that both will occur. In everyday communication, people generally rely on context to make their intended meaning clear. But in propositional logic, there is no context, only the bare knowledge of truth or falsity. Since mathematics requires unambiguous definitions, mathematicians had to make a choice when they were formulating the rules of propositional logic, and they chose the inclusive version. Their decision is reflected in the truth table for disjunction. Since it is an easy matter to express the exclusive-or in terms of the inclusive-or and the other logical operations, there is no net loss in making this particular choice. But the choice was not arbitrary. Mathematicians chose the inclusive-or because it leads to a logical pattern much more similar to that of Boolean algebra, described in the previous section.

There is no common English word that directly corresponds to the conditional operation. It is related to logical implication, so "implies" would be the closest word. But the conditional does not really capture the notion of implication. Implication involves some kind of causality; if I say that *p implies q* (an alternative way to say the same thing is *If p, then q*), you will understand there to be some sort of connection between *p* and *q*. But the operations of propositional logic are defined solely in terms of truth and falsity, and this method of definition is too narrow to capture the notion of implication. The conditional does its best, by capturing the two patterns of truth that arise from an implication, namely:

- If it is the case that *p* implies *q*, then the truth of *q* follows from that of *p*.
- If it is the case that *p* is true and *q* is false, then it cannot be the case that *p* implies *q*.

These considerations give you the first two rows of the truth table for the conditional. The remainder of the table, which concerns the two cases in which *p* is false, is completed in a fashion that leads to the most useful theory. Here you have an example of being guided by a *mathematical* pattern when there is no real-world pattern to go by.

Since the various logical operations are defined purely in terms of their truth patterns, if two compound propositions have truth tables that are row-by-row identical, then the two compounds are, to all intents and purposes, equal. By computing truth tables, we can derive various laws of logical algebra. By using the symbol $\otimes$ to stand for *and*, the symbol $\oplus$ to stand for *or*, and a minus sign to denote negation, you will be able to see the ways in which these logical connectives resemble, respectively, the arithmetic operations $\times$, $+$, and $-$, and the ways in which they differ. To make the comparison with arithmetic even clearer, I'll use 1 to stand for any true proposition (such as $5 = 5$) and 0 for any false proposition (such as $5 = 6$).

$$p \otimes q = q \otimes p$$
$$p \otimes (q \otimes r) = (p \otimes q) \otimes r$$
$$p \otimes (q \oplus r) = (p \otimes q) \oplus (p \otimes r)$$
$$p \otimes 1 = p$$
$$p \otimes 0 = 0$$
$$-(p \otimes q) = (-p) \oplus (-q)$$
$$-(-p) = p$$
$$p \to q = (-p) \oplus q$$

$$p \oplus q = q \oplus p$$
$$p \oplus (q \oplus r) = (p \oplus q) \oplus r$$
$$p \oplus (q \otimes r) = (p \oplus q) \otimes (p \oplus r)$$
$$p \oplus 1 = 1$$
$$p \oplus 0 = p$$
$$-(p \oplus q) = (-p) \otimes (-q)$$

A much stronger connection unites the above patterns and those in Boole's algebra. If $p \otimes q$ is taken to correspond to Boole's product pq, $p \oplus q$ to Boole's sum $p + q$, and $-p$ to Boole's $1 - p$, and if 1, 0 are taken to correspond to Boole's 1, 0, respectively, then all of the above identities (apart from the last one) hold for Boole's logic.

In all of these identities, the 'equality' is not genuine equality. All it means is that the two propositions concerned have the same truth table. In particular, with this meaning of 'equals', the following is true:

7 is a prime = the angles of a triangle sum to 180°.

Because of this special meaning of 'equality', mathematicians generally use a different symbol, writing $\leftrightarrow$ or $\equiv$ instead of = in such identities.

Incidentally, though many of the logical properties expressed by the algebraic equations listed above were discovered by the Stoics, they did not use algebraic notation at all; they wrote everything out in words. As you can imagine, this meant that they had to deal with long, practically unreadable sentences. This was almost certainly the reason why Stoic logic remained dormant and unstudied until Boole came along and showed how to use algebraic notation to study logical patterns.

The patterns of reason

Patterns of truth explain the rules for combining propositions. But what are the patterns involved in deducing one proposition from another? Specifically, in propositional logic, what takes the place of Aristotle's syllogisms?

The answer is the following simple deduction rule, known to the Stoics, who called it *modus ponens*:

From $p \rightarrow q$ and p, infer q.

This rule clearly accords with the intuition that the conditional corresponds to the notion of implication.

It should be stressed that the p and q here do not have to be simple, noncompound propositions. As far as modus ponens is concerned, these symbols may denote any proposition whatsoever. Indeed, throughout propositional logic, the algebraic symbols used almost invariably denote arbitrary propositions, simple or compound.

In propositional logic, a proof, or valid deduction, consists of a series of propositions such that each proposition in the series is either deduced from previous ones by means of modus ponens or is one of the assumptions that underlie the proof. In the course of a proof, any of the logical identities listed in the previous section may be used, just as any of the laws of arithmetic may be used in the course of a calculation.

Though it does not capture all kinds of reasoning, not even all kinds of mathematical proof, propositional logic has proved to be extremely useful. In particular, today's electronic computer is, to all intents and purposes, simply a device that can perform deductions in propositional logic. Indeed, the two great pioneers of computing, Alan Turing and John von Neumann, were both specialists in mathematical logic.

Splitting the logical atom

The final step in trying to capture the patterns involved in mathematical proofs was supplied by Guiseppe Peano and Gottlob Frege at the end of the nineteenth century. Their idea was to 'split the logical atom'; that is, they showed how to break apart the basic propositions that propositional logic takes as unanalyzed, atomic entities. Specifically, they took propositional logic and added further deductive mechanisms that depend on the nature of the propositions, not just their truth values. In a sense, their system of logic combined the strengths of both Aristotle's approach—in which the rules of deduction do depend on the nature of the propositions—and propositional logic, which captures deductive patterns of pure logic. However, the additional rules used to produce *predicate logic*, as the new theory is known, are far more general than those for Aristotle's syllogisms (though they do involve the notions of 'all' and 'some' that formed the basis of the figures of the syllogism).

Predicate logic has no unanalyzed, atomic propositions. All propositions are regarded as built up from more basic elements. In other words, in predicate logic, the study of the patterns of deduction is preceded by, and depends upon, the study of certain linguistic patterns— the patterns of language used to form propositions.

This system of logic takes as its basic elements not propositions, but *properties*, or *predicates*. The simplest of these are the very same predicates found in Aristotle's logic, such as:

... *is a man.*

... *is mortal.*

... *is an Aristotle.*

However, predicate logic allows for more complex predicates, involving two or more objects, such as

... *is married to* ... ,

which relates two objects (people), or

> *. . . is the sum of . . . and . . . ,*

which relates three objects (numbers).

Predicate logic extends propositional logic, but the focus shifts from propositions to sentences (the technical term is *formulas*). This shift in focus is necessary because predicate logic allows you to construct sentences that are not necessarily true or false, and thus do not represent propositions. The construction rules comprise the propositional operations *and, or, not,* and the conditional (→), plus the two quantifiers *all* and *some.* As in Aristotle's logic, *some* is taken to mean 'at least one', as in the sentence *Some even number is prime.* Another phrase that means the same is *There exists,* as in *There exists an even prime number.*

The actual rules for sentence construction—the *grammar* of predicate logic—are a bit complicated to write down precisely and completely, but the following simple examples should give the general idea.

In predicate logic, Aristotle's proposition *All men are mortal* is constructed like this:

> *For all x, if x is a man, then x is mortal.*

This construction looks more complicated than the original version, and it is certainly more of a mouthful to say. The gain is that the proposition has been pulled apart into its constituents, exposing its internal, logical structure. This structure becomes more apparent when the logician's symbols are used instead of English words and phrases.

First, the logician writes the predicate *x is a man* in the abbreviated form Man(x) and the predicate *x is mortal* in the form Mortal(x). This change of notation may, on occasion, make something simple appear complex and mysterious, but that is certainly not the intention. Rather, the aim is to direct attention to the important patterns involved. The crucial aspect of a predicate is that it is true or false of one or more particular objects. What counts are (1) the property and (2) the objects; everything else is irrelevant.

The proposition *Aristotle is a man* would thus be written

> Man(Aristotle).

The proposition *Aristotle is not a Roman* would be written

> *not*-Roman(Aristotle).

The proposition *Susan is married to Bill* would be written

> Married-to(Susan, Bill).

This notation highlights the general pattern that a predicate is something that is true (or not) of certain objects:

Predicate(object, object, ...)

not-Predicate(object, object, ...)

Two further symbols may be used. The word *All* or the phrase *For all* can be abbreviated by an upside-down letter A: $\forall$. The word *Some* or the phrase *There exists* can be abbreviated by a back-to-front letter E: $\exists$. Using this notation, *All men are mortal* looks like this:

$\forall x : \text{Man}(x) \rightarrow \text{Mortal}(x).$

Written in this way, all the logical constituents of the proposition and the underlying logical pattern are immediately obvious:

- the quantifier, $\forall$;
- the predicates, Man and Mortal;
- the logical connection between the predicates, namely, $\rightarrow$.

As a final example, the proposition *There is a man who is not asleep* looks like this:

$\exists x : \text{Man}(x)$ *and* not-$\text{Asleep}(x).$

Though propositions written in this way look strange to anyone who sees them for the first time, logicians have found this notation to be extremely valuable. Moreover, predicate logic is powerful enough to express all mathematical propositions. It is certainly true that a definition or proposition expressed in predicate logic can appear daunting to the uninitiated. But this is because such an expression does not hide any of the logical structure; the complexity you see in the expression is the actual structural complexity of the notion defined or of the proposition expressed.

As with the operations of propositional logic, there are rules that describe the algebraic properties of the operations of predicate logic. For example, there is the rule

not-$[\forall x : P(x)] \equiv \exists x : not$-$P(x).$

In this rule, $P(x)$ can be any predicate that is applicable to a single object, such as $\text{Mortal}(x)$. Written in English, an instance of this rule would be

Not all men like football $\equiv$ *Some men do not like football.*

Again, just as was the case with the rules for propositional logic, the symbol ≡ means that the two expressions 'say the same thing'.

The development of predicate logic provided mathematicians with a means to capture, in a formal way, the patterns of mathematical proof. This is not to say that anyone has ever advocated a slavish adherence to the rules of predicate logic. No one insists that all mathematical assertions be expressed in predicate logic, or that all proofs be formulated only in terms of modus ponens and the deduction rules that involve quantifiers (not stated here). To do so would, except for the simplest of proofs, be extremely laborious, and the resulting proof would be almost impossible to follow. But, by virtue of carrying out a detailed study of the patterns of predicate logic, mathematicians not only gained considerable understanding of the concept of a formal proof, they also confirmed that it was a reliable means of establishing mathematical truth. And that was of the greatest importance in view of other developments taking place in mathematics at the same time, to which I turn next.

The dawn of the modern age

The second half of the nineteenth century was a glorious era of mathematical activity. In particular, it was during this period that mathematicians finally managed to work out a proper theory of the real-number continuum, thereby providing a rigorous foundation for the methods of the calculus that Newton and Leibniz had developed three hundred years earlier (see Chapter 3). Crucial to this progress was the increasing—and in many cases total—reliance on the axiomatic method.

All mathematics deals with abstraction. Though many parts of mathematics are motivated by, and can be used to describe, the physical world, the entities the mathematician actually deals with—the numbers, the geometric figures, the various patterns and structures—are pure abstractions. In the case of subjects such as the calculus, many of these abstractions involve the mathematical notion of infinity and thus cannot possibly correspond directly to anything in the real world.

How does the mathematician decide whether some assertion about these abstractions is true or not? The physicist or chemist or biologist generally accepts or rejects a hypothesis on the basis of experiment, but most of the time the mathematician does not have this option. In cases that can be settled by straightforward numerical computation, there is no problem. But in general, observations of events in the real world are at best suggestive of some mathematical fact, and on occasion can be downright misleading, with the mathematical truth being quite at variance with everyday experience and intuition.

The existence of nonrational real numbers falls into this category of counterintuitive mathematical facts. Between every two rational numbers lies a third: namely, the mean of the first two. On the basis of everyday experience, it would therefore seem reasonable to suppose that there is simply no room on the rational line for any more numbers. But as the Pythagoreans discovered, to their immense dismay, this is not the case at all.

Although the Pythagoreans were devastated by their discovery, subsequent mathematicians accepted the existence of irrational numbers because their existence had been *proved*. Ever since Thales, proof has been central to mathematics. In mathematics, truth is not decided by experiment, nor by majority vote, nor by dictate—even if the dictator is the most highly regarded mathematician in the world. Mathematical truth is decided by proof.

This is not to say that proofs are all there is to mathematics. As the science of patterns, a lot of mathematics is concerned with finding new patterns in the world, analyzing those patterns, formulating rules to describe them and facilitate their further study, looking for their appearance in a new domain of patterns observed somewhere else, and applying theories and results to phenomena in the everyday world. In many of these activities, a reasonable question is, how well do these mathematical patterns and results accord with what may be physically observed, or with what may be computed? But as far as establishing mathematical truth is concerned, there is only one game in town: proof.

Mathematical truths are all fundamentally of the form

If A, then B.

In other words, all mathematical facts are proved by deduction from some initial sets of assumptions, or *axioms* (from the Latin *axioma*, meaning 'a principle'). When a mathematician says that a certain fact B is "true," what she means is that B has been proved on the basis of some set of assumed axioms A. It is permissible to express this result simply as "B is true," provided the axioms A are obvious, or at least widely accepted within the mathematical community.

For example, all mathematicians would agree that it is true that, between any positive whole number N and its double $2N$, there is a prime number. How can they be so sure? They certainly have not examined every possible case, since there are an infinite number of cases. Rather, this result has been *proved*. Moreover, the proof depends only on a set of axioms for the natural numbers that everyone accepts as definitive.

Provided you are sure that a proof is valid, the only part of this process that remains open to question is whether or not the axioms correspond to your intuitions. Once you write down a set of axioms, then anything you prove from those axioms will be mathematically true for the system of objects your axioms describe. (Strictly speaking, I should say "*any* system of objects," since most axiom systems will describe more than one system of objects, regardless of the purpose for which the axioms were originally formulated.) But it might well be that the system your axioms describe is not the one you set out to describe.

For instance, around 350 B.C., Euclid wrote down a set of axioms for the plane geometry of the world around us. From this set of axioms he was able to prove a great many results, results both aesthetically pleasing and immensely useful in everyday life. But in the nineteenth century, it was discovered that the geometry described by Euclid's axioms might not be the geometry of the world around us after all. It might be only approximately right, albeit with a degree of approximation that is not noticeable in everyday life. In fact, present-day theories of physics assume geometries different from Euclid's. (This fascinating story is fully described in Chapter 4.)

By way of illustration of the axiomatic method, here is a set of axioms formulated during the nineteenth century for the elementary arithmetic of the *integers*, the positive and negative whole numbers.

1. For all m, n: $m + n = n + m$ and $nm = mn$ (the commutative laws for addition and multiplication).
2. For all m, n, k: $m + (n + k) = (m + n) + k$ and $m(nk) = (mn)k$ (the associative laws for addition and multiplication).
3. For all m, n, k: $k(m + n) = (km) + (kn)$ (the distributive law).
4. For all n: $n + 0 = n$ (the additive identity law).
5. For all n: $1n = n$ (the multiplicative identity law).
6. For all n, there is a number k such that $n + k = 0$ (the additive inverse law).
7. For all m, n, k, where $k \neq 0$: if $km = kn$, then $m = n$ (the cancellation law).

These axioms are widely accepted by mathematicians as describing the arithmetic of the integers. In particular, anything proved on the basis of these axioms will be described by a mathematician as 'true'. And yet, it is easy to write down 'facts' that no one has any hope of checking, either by direct computation or by experimental processes such as counting piles of pennies. For instance, is the following identity true?

$$12{,}345^{678{,}910} + 314{,}159^{987{,}654{,}321} = 314{,}159^{987{,}654{,}321} + 12{,}345^{678{,}910}$$

This identity is of the form $m + n = n + m$, so, on the basis of the axioms, you know it is 'true'. (In fact, the commutative law for addition *says* it is true; you don't even have to construct a proof in this case.) Is this a reliable way to 'know' something?

In writing down the above axioms for integer arithmetic, the mathematician is describing certain patterns that have been observed. Everyday experience with small numbers tells you that the order in which you add or multiply two numbers does not affect the answer. For example, $3 + 8 = 8 + 3$, as you can show by counting out coins. If you count out three coins and then add a further eight, the number of coins you end up with is the same as if you first count out eight coins and then add a further three; in both cases you end up with eleven coins. This pattern is repeated with every pair of numbers you encounter. What is more, it seems reasonable to *assume* that the pattern will continue to be true for any other pair of numbers you might encounter tomorrow, or the next day, or even for any pair of numbers anyone else might encounter at any time in the future. The mathematician takes this reasonable assumption, based on everyday experience, and *declares* it to be 'true' for *all* pairs of integers, positive or negative.

Because such rules are taken as axioms, any collection of objects that obeys all the rules listed above will have any property that can be proved on the basis of those axioms. For example, using the above axioms, it is possible to prove that the additive inverse of any 'number' is unique; that is, for any 'number' n, the 'number' k such that $n + k = 0$ is unique. Thus, whenever you have a system of 'numbers' that satisfies all these axioms, that system will never contain a 'number' that has two additive inverses.

The reason for the quotes around the word 'number' in the above paragraph is that, as mentioned earlier, whenever you write down a set of axioms, it generally turns out that there is more than one system of objects that satisfies those axioms. Systems that satisfy the above axioms for integer arithmetic are called *integral domains.*

Mathematicians have encountered a number of objects besides integers that form integral domains. For example, polynomial expressions form an integral domain. So too do certain of the finite arithmetics described in Chapter 1. In fact, the finite arithmetics with a prime number modulus satisfy an additional, eighth axiom beyond those for an integral domain, namely:

8. For all n other than 0, there is a k such that $nk = 1$.

This is the multiplicative inverse law. It implies Axiom 7, the cancellation law. (More precisely, on the assumption of Axioms 1 through 6 and Axiom 8, it is possible to prove Axiom 7.) Systems that satisfy Axioms 1 through 8 are called *fields.*

There are many examples of fields in mathematics. The rational numbers, the real numbers, and the complex numbers are all fields (see Chapter 3). There are also important examples of fields in which the objects are not 'numbers' in the sense in which most people would understand this word.

The power of abstraction

To the person who meets modern abstract mathematics for the first time, it might all seem like a frivolous game—not that there is anything wrong with frivolous games! But the formulation of axioms and the deduction of various consequences of those axioms has, over the years, proved to be an extremely powerful approach to many kinds of phenomena, with direct consequences for everyday life, both good and ill. Indeed, most of the components of modern life are based on the knowledge humankind has acquired with the aid of the axiomatic method. (That's not *all* they are based on, of course, but it is an essential component. Without the axiomatic method, technology, for instance, would have advanced little beyond that of a century ago.)

Why has the axiomatic method been so successful? The answer is, in large part, because the axioms used do indeed capture meaningful and correct patterns.

Which statements are accepted as axioms often depends as much on human judgment as anything else. For instance, on the basis of embarrassingly little concrete evidence, most citizens would be prepared to stake their lives on the validity of the commutative law of addition. (How many times in your life have you bothered to check this law for a particular pair of numbers? Think of this the next time you step onto an airplane, where your life literally does depend on it!)

There is certainly no logical basis for this act of faith. Mathematics abounds with examples of statements about numbers that are true for millions of cases, but not true in general. The Mertens conjecture, for example, is a statement about natural numbers that had been verified by computer to be true for the first 7.8 billion natural numbers before it was proved false in 1983. Yet, even before it was proved false, no one had ever suggested adding this statement to the axioms for the natural numbers.

Why have mathematicians adopted the commutative law, for which the numerical evidence is slim, as an axiom and left out other assertions for which there is a huge amount of numerical evidence? The decision is essentially a judgment call. For a mathematician to adopt a certain pattern as an axiom, that pattern not only should be a useful assumption but also should be 'believable', in keeping with her intuitions, and as simple as possible. Compared with these factors, supporting numerical evidence is of relatively minor importance—though a single piece of numerical evidence to the contrary would at once overturn the axiom!

Of course, there is nothing to prevent anyone from writing down some arbitrary list of postulates and proceeding to prove theorems from them. But the chances of those theorems having any practical application, or even any application elsewhere in mathematics, are slim indeed. And such an activity would be unlikely to find acceptance within the mathematical community. Mathematicians are not too disturbed when their work is described as "playing games." But they become really annoyed if it is described as "meaningless games." And the history of civilization is very much on the mathematicians' side: there is usually no shortage of applications for their results.

The reason why the mathematician seeks the comfort of starting with a set of *believable* axioms for some system is that, once he starts trying to understand the system by proving consequences of those axioms, everything he does will rest upon those initial axioms. The axioms are like the foundations of a building. No matter how carefully the mathematician constructs the walls and the rest of the structure, if the foundations are unsound, the entire structure may collapse. One false axiom, and everything that follows may be wrong or meaningless.

As outlined earlier, the initial step in the development of a new branch of mathematics is the identification of some pattern. Then comes the abstraction of that pattern to a mathematical object or structure, say, the concept of a natural number or a triangle. As mathematicians study that abstract concept, the various patterns they observe might lead to the formulation of axioms. At that point, there is no longer any need to know about the phenomenon that led to those axioms in the first place. Once the axioms are available, everything can proceed on the basis of logical proofs, carried out in a purely abstract setting.

Of course, the pattern that starts this whole process may be something that arises in the everyday world; for example, this is the case for the patterns studied in Euclidean geometry and, to some extent, in elementary number theory. But it is quite possible to take patterns that themselves arise from mathematics and subject *those* patterns to the same process of abstraction. In this case, the result will be a new level of abstraction. The definition of integral domains is an example of this process of higher-level abstraction. The axioms for an integral domain capture a pattern that is exhibited not only by the integers, but also by polynomials and a number of other mathematical systems, each of which is itself an abstraction capturing other, lower-level patterns.

During the nineteenth century, this process of abstracting from abstractions was taken to an extent that left all but a relatively small handful of professional mathematicians unable to appreciate most of the new developments in mathematics. Abstractions were piled on abstractions to form an enormous tower, a process that continues to this day. Though high levels of abstraction may cause many people to shy away from modern mathematics, increasing the level of abstraction does not in itself lead to more difficult mathematics. At each level of abstraction, the actual mechanics of doing mathematics remain much the same. Only the level of abstraction changes.

It is interesting to note that the trend toward increased abstraction over the past hundred years or so has not been unique to mathematics. The same process has been taking place in literature, in music, and in the visual arts—often resulting in a similar lack of appreciation by those not directly involved in the process.

The versatile concept of sets

As the level of abstraction in mathematics increased, mathematicians grew ever more dependent on the notion of abstract *sets* ('set' being the name they adopted as a technical term to refer to any collection of objects of some kind).

New mathematical notions, such as groups, integral domains, fields, topological spaces, and vector spaces, were being introduced and studied, and many of these were defined as sets of objects on which certain operations could be performed (operations such as 'addition' and 'multiplication' of various kinds). Familiar old notions from geometry, such as lines, circles, triangles, planes, cubes, octahedra, and the like, were given new definitions as sets of points satisfying various conditions. And of course, Boole had developed his algebraic treatment of logic by regarding the syllogism in terms of sets.

The first complete mathematical theory of abstract sets was worked out by the German mathematician Georg Cantor toward the very end of the nineteenth century, although the beginnings of this theory were clearly present in the work of Boole. The basic ideas of Cantor's theory can be found in Boole's treatment of the syllogism. The theory starts by developing an 'arithmetic' for sets:

> If x and y are sets, let xy denote the set consisting of all members common to x and y, and let $x + y$ denote the set consisting of all members of x together with all members of y.

The only difference between this definition and the one given earlier for Boole's logic is that the symbols x and y are now regarded as standing for *any* sets, not just sets arising from a logical proposition. The following 'arithmetical' axioms, given previously for Boole's classes, are true in this more general situation:

$$x + y = y + x \qquad\qquad xy = yx$$
$$x + (y + z) = (x + y) + z \qquad x(yz) = (xy)z$$
$$x(y + z) = xy + xz$$
$$x + 0 = x$$
$$x + x = x \qquad\qquad\qquad xx = x$$

(Boole's axiom involving the object 1 does not appear, since there is no need for such an object in set theory, and introducing one can lead to technical problems.)

In present-day set theory, the set xy is called the *intersection* of x and y, and the set $x + y$ is called the *union*. A more common notation for these operations is $x \cap y$ for the intersection and $x \cup y$ for the union. In addition, contemporary mathematicians usually denote the empty set—the set having no members—by $\varnothing$ rather than 0. (The empty set is to set theory what the number zero is to arithmetic.)

For small sets, with no more than a dozen or so members, mathematicians use a notation that explicitly lists the members. For example, the set consisting of the numbers 1, 3, and 11 is written like this:

$$\{1, 3, 11\}.$$

Larger sets, or infinite ones, clearly cannot be depicted in this manner; in those cases, other ways have to be found to describe the set. If the members of a set have an obvious pattern, that pattern can be used to describe the set. For example:

$$\{2, 4, 6, \ldots\}$$

denotes the infinite set of all even natural numbers. Often, the only reasonable way to describe the set is in words, such as "the set of all primes."

It is common to write $x \in A$ to denote that the object x is a member of the set A, and $x \notin A$ to denote that x is not a member of A.

Numbers from nothing

For all its apparent simplicity, set theory proved to be extremely powerful. Mathematicians even managed to use set theory to answer that most fundamental of questions: what exactly are numbers?

Of course, most of the time even mathematicians don't ask themselves this question. They simply use numbers, just like everybody else. But in science, there is always a desire to reduce any concept to something simpler and more basic. So too in mathematics.

To answer the question 'what is a number?' the mathematician shows how the real numbers can be described in terms of rational numbers, the rational numbers in turn described in terms of integers, the integers in terms of natural numbers, and the natural numbers in terms of sets. Without going into any of the details, here are the first three of these reductions:

Real numbers can be defined as certain pairs of certain (infinite) sets of rational numbers.

Rational numbers can be defined as certain (infinite) sets of pairs of integers.

Integers can be defined as pairs of natural numbers.

Provided you accept the notion of an arbitrary set as basic, the unraveling process comes to an end by providing a description of the natural numbers. Moreover, it does so in a surprising way. In set theory, it is possible to construct the entire infinitude of natural numbers starting with nothing—more precisely, with the empty set, $\emptyset$. The procedure goes like this.

Define the number 0 *to be* the empty set, $\emptyset$.

The number 1 is then defined to be the set {0}, the set having exactly one member, that member being the number 0. (If you unravel this step, you find that the number 1 is equal to the set {$\emptyset$}. If you think about it for a moment, you will realize that this is not quite the same as the empty set; $\emptyset$ has no members, whereas the set {$\emptyset$} has one member.)

The number 2 is defined to be the set {0,1}. The number 3 is the set {0,1,2}. And so forth.

Each time a new number is defined, you use it and all the previous numbers to define the next one. In general, the natural number n is the set consisting of 0 and all natural numbers smaller than n:

$$n = \{0, 1, 2, \ldots, n - 1\}.$$

(So the natural number n is a set with exactly n members.) Notice that the whole process starts out from the empty set, $\emptyset$ — that is to say, from 'nothing'. Very clever.

Cracks in the foundations

By the turn of the century, set theory, its power recognized by all, had become a general framework for an extremely large part of mathematics. So it was with considerable alarm that the world of mathematics woke up one June morning in 1902 to find that set theory was fundamentally inconsistent—that is, in Cantor's set theory, it was possible to prove that 0 equals 1. (Strictly speaking, the problem arose in an axiomatization of set theory by the Swedish mathematician Gottlob Frege. But Frege's axioms simply formalized Cantor's ideas.)

Of all the things that can be wrong with an axiom system, inconsistency is absolutely the worst. It is possible to work with axioms that are hard to understand. It is possible to work with axioms that are counterintuitive. And all might not be lost if your axioms don't accurately describe the structure you intended to capture—maybe they will find some other application, as has happened on more than one occasion. But an inconsistent set of axioms is completely useless.

The inconsistency was found by Bertrand Russell, just as the second volume of Frege's treatise on his new theory was at the printer. Russell's argument was devastatingly simple.

According to Cantor and Frege, and most likely to any mathematician at the time who cared to think about it, for any property P, there will surely be a corresponding set of all objects that have the property P. For example, if P is the property of being a triangle, then the set corresponding to P is the set of all triangles. (Frege's work amounted in large part to developing a formal theory of properties to go with this idea; that theory was the predicate logic explored earlier in the chapter.)

The standard notation for the set of all objects x that have the property P is

$$\{x \mid P\}.$$

For example, the set of all prime numbers can be expressed as

$$\{x \mid x \text{ is a prime number}\}.$$

In this example, the property P applies to natural numbers, and the resulting set is a set of natural numbers. Russell's argument concerns properties P that apply to sets; for such a property, the corresponding set is a set of sets.

One property P that applies to sets is that a set be a member of itself. Some sets do have this property; they are members of themselves. For example, if M denotes the set of all sets that are explicitly named in this book, then M is a member of itself ($M \in M$). On the other hand, there are sets that do not have the property P; these sets are not members of themselves. For example, the set of all natural numbers ($\mathcal{N}$) is not itself a natural number, and hence is not a member of itself ($\mathcal{N} \notin \mathcal{N}$).

To arrive at a contradiction, Russell looked not at the property P, but at the closely related property R that a set x *not* be a member of itself. Some sets do have this property (for example, $\mathcal{N}$); others do not (M). Since R is a perfectly reasonable-looking property, albeit a bit novel, according to Cantor and Frege there should be a corresponding set—call it $\mathcal{C}$. $\mathcal{C}$ is the set of all those sets that have the property R. In symbols:

$$\mathcal{C} = \{x \mid x \text{ is a set and } x \notin x\}.$$

So far so good. But Russell now asks the question, again perfectly reasonable: is this new set $\mathcal{C}$ a member of itself or is it not?

Russell points out that if $\mathcal{C}$ is a member of itself, then $\mathcal{C}$ must have the property R that defines $\mathcal{C}$. And what that means is that $\mathcal{C}$ is *not* a member of itself. So $\mathcal{C}$ both is and is not a member of itself, which is an impossible situation.

Russell continues by asking what happens if $\mathcal{C}$ is *not* a member of itself. In that case, $\mathcal{C}$ must fail to satisfy the property R. Therefore, it is not the case that $\mathcal{C}$ is not a member of itself. That last clause is just a complicated way of saying that $\mathcal{C}$ is a member of itself. So again, the inescapable conclusion is that $\mathcal{C}$ both is not and is a member of itself, an impossible state of affairs.

And now you are at a complete impasse. Either $\mathcal{C}$ is a member of itself or it is not. Either way, you finish by concluding that it both is and is not. This result is known as Russell's paradox. Its discovery indicated that something was wrong with Cantor's set theory—but what?

Since Russell's reasoning was correct, there seemed to be only one way to resolve the paradox: the definition of the set $\mathcal{C}$ must be faulty in some way. And yet, it is just about as simple a definition of a set as anyone could hope for. The sets used to construct the various number systems are far more complicated. Unpalatable as it might have been, therefore, mathematicians had to jettison the assumption that for any property there will be a corresponding set—the set of objects having that property.

The situation was not unlike that facing the Pythagoreans when they discovered that there were lengths that did not correspond to any known numbers. And once again, there was no choice in the matter. Faced with a proof that something is fundamentally wrong, a theory has to be modified or replaced, regardless of how simple, elegant, or intuitive it is. Cantor's set theory had been all three: simple, elegant, and intuitive. But it had to be abandoned.

What replaced Cantor's set theory was an *axiomatic* theory of sets developed by Ernst Zermelo and Abraham Fraenkel. Though Zermelo–Fraenkel set theory manages to remain close in spirit to Cantor's highly intuitive notion of an abstract set, and though it has proved to be an adequate foundation for all of pure mathematics, it has to be admitted that the theory is not particularly elegant. Compared with Cantor's theory, the seven axioms introduced by Zermelo and the subtle additional axiom introduced by Fraenkel form something of a motley collection. They describe rules that give rise to the various sets required in mathematics while carefully skirting the kind of difficulty that Russell had uncovered.

Zermelo and Fraenkel's analysis of sets is sufficiently compelling that most mathematicians accept the resulting axioms as the right ones on which to base mathematics. Yet for many, their first encounter with Russell's paradox and the steps required to circumvent it produces a sense of innocence lost. For all that the concept of a pure set might seem the very essence of simplicity, closer analysis reveals otherwise. Set theory may well be an ultimate pure creation of the human intellect, the essence of abstraction, but, as is the case with all great constructions of mathematics, it dictates its own properties.

The rise and fall of Hilbert's program

Thirty years after Russell destroyed Cantor's intuitive set theory, a similar upheaval occurred, with equally devastating consequences. The victim on this second occasion was the axiomatic method itself, which had found its most influential champion in the German mathematician David Hilbert.

The axiomatic approach to mathematics made it possible for mathematicians to separate the issues of provability and truth. A mathematical proposition was *provable* if you could find a logically sound argument that deduced it from the appropriate axioms. A proposition that had been proved was *true* provided the assumed axioms were true. The former notion, provability, was a purely technical one in which the mathematician ruled supreme; the latter notion, truth, involved deep philosophical questions. By separating these two notions, mathematicians could sidestep thorny questions about the nature of truth and concentrate on the issue of proof. By restricting themselves to the task of proving results on the basis of an assumed axiom system, mathematicians were able to regard mathematics as a formal game—a game played according to the rules of logic, starting from the relevant axioms.

Discovering the appropriate axioms was clearly an important component of this formalistic approach to mathematics, as it became known. Implicit in formalism was an assumption that, provided you looked long enough, you would eventually find all the axioms you needed. In this way, the *completeness* of the axiom system became a significant issue: had enough axioms been found to be able to answer all questions? In the case of the natural numbers, for example, there was already an axiom system, formulated by Peano. Was this axiom system complete, or were additional axioms needed?

A second important question was, is the axiom system consistent? As Russell's paradox demonstrated all too clearly, writing down axioms that describe a highly abstract piece of mathematics is a difficult task.

The purely formalistic approach to mathematics, involving as it did the search for consistent, complete axiom systems, became known as the Hilbert program, named after David Hilbert, one of the leading mathematicians of the time. Though he was not a logician in the sense that Frege and Russell were, questions about the foundations of mathematics were particularly important to Hilbert, whose own work was of a highly abstract nature. For instance, one of his legacies to mathematics is the Hilbert space, a sort of infinite-dimensional analogue of three-dimensional Euclidean space.

Any dreams that the Hilbert program could be completed were dashed in 1931, when a young Austrian mathematician named Kurt Gödel (see Figure 2.3) proved a result that was to change our view of mathematics forever. Gödel's theorem says that if you write down any consistent axiom system for some reasonably large part of mathematics,

FIGURE **2.3** Kurt Gödel (1906–1978).

then that axiom system must be incomplete: there will always be some questions that cannot be answered on the basis of the axioms.

The phrase 'reasonably large' in the above paragraph is meant to exclude trivial examples such as the geometry of a single point (0-dimensional geometry). In order to prove Gödel's theorem, you need to know that the axiom system concerned includes, or is rich enough to allow the deduction of, all the axioms of elementary arithmetic. This is probably as weak a requirement to put on a system of axioms for mathematics as you will ever see.

The proof of Gödel's totally unexpected result is highly technical, but his idea is a simple one, having its origins in the ancient Greek paradox of the liar. Suppose a person stands up and says, "I am lying." If this assertion is true, then the person is indeed lying, which means that what he says is false. On the other hand, if the assertion is false, then the person must not be lying, so what he says is true. Either way the assertion is contradictory.

Gödel found a way to translate this paradox into mathematics, replacing truth with provability. His first move was to show how to translate propositional logic into number theory. Included in this translation process was the notion of a formal proof from the axioms. He then produced a particular number-theoretic proposition that says, in effect,

(*) *The proposition on this page labeled with an asterisk is not provable.*

First of all, proposition (*) (or rather, Gödel's formal version of it) must be either true or false (for the mathematical structure under consideration—say, the arithmetic of natural numbers, or set theory). If the proposition is false, then it must be provable. You can see that by just looking at what (*) says. But, since the consistency of the axiom system is assumed, anything that is provable must be true. So, if (*) is false, then it is true, which is an impossible situation. Hence (*) must be true.

Is proposition (*) provable (from the axioms concerned)? If it is, then, as mentioned a moment ago, it must be true for that system, and that means it is not provable. This is a contradictory state of affairs. The conclusion is that proposition (*) cannot be proved from the given axioms. So (*) is a proposition that is true for the structure, but not provable from the axioms for that structure.

Gödel's argument can be carried through for any set of axioms you can write down for a mathematical structure. The stipulation that you must be able to write down the axioms is important. After all, there is one trivial way to obtain an axiom system that can be used to prove all true propositions about the structure, and that is to declare the set of all true propositions *to be* the set of axioms. This is clearly not in the spirit of the axiomatic approach to mathematics, and is a quite useless axiomatization.

On the other hand, the phrase 'write down' can be taken in a very broad, idealistic sense. It allows not only for large finite sets of axioms, which can be written down only in principle, but also for infinite sets of axioms of certain kinds. The key requirement is that you be able to stipulate one or more rules that show how the axioms *could* be written down. In other words, the axioms themselves must exhibit a very definite linguistic pattern. Peano's axioms for the natural numbers and the Zermelo–Fraenkel axioms for set theory are both infinite axiom systems of this kind.

Gödel's discovery that, for important areas of mathematics such as number theory or set theory, no consistent set of axioms can be complete clearly makes it impossible to achieve the goal of Hilbert's program. In fact, the situation is even worse: Gödel went on to show that among the propositions that are true but unprovable from the axioms is one that asserts the consistency of those axioms. So there is not even the hope of proving that your axioms are consistent.

In short, in the axiomatization game, the best you can do is to *assume* the consistency of your axioms and *hope* that they are rich enough to enable you to solve the problems of highest concern to you. You have to accept that you will be unable to solve all problems using your axioms; there will always be true propositions that you cannot prove from those axioms.

The golden age of logic

Though it marked the end of the Hilbert program, the proof of Gödel's theorem ushered in what can only be described as the Golden Age of logic. The period from around 1930 through the late 1970s saw intense activity in the area that became known by the general term 'mathematical logic'.

From its very beginning, mathematical logic split into several, connected strands.

Proof theory took to new lengths the study of mathematical proofs begun by Aristotle and continued by Boole. In recent years, methods and results from this branch of mathematical logic have found uses in computing, particularly in artificial intelligence.

Model theory, invented by the Polish-born American mathematician Alfred Tarski and others, investigated the connection between truth in a mathematical structure and propositions about that structure. The result alluded to earlier, that any axiom system will be true of more than one structure, is a theorem of model theory. In the 1950s, the American logician and applied mathematician Abraham Robinson used techniques of model theory to work out a rigorous theory of infinitesimals, thereby providing a way to develop the calculus (see Chapter 3) quite different from the one worked out in the nineteenth century, and in many ways superior to it.

Set theory took on new impetus when model-theoretic techniques were brought to bear on the study of Zermelo–Fraenkel set theory. A major breakthrough came in 1963, when Paul Cohen, a young American mathematician, found a means to prove, rigorously, that certain mathematical statements were *undecidable*—that is to say, could be proved neither true nor false from the Zermelo–Fraenkel axioms. This result was far more wide-ranging than Gödel's theorem. Gödel's theorem simply tells you that, for an axiom system such as that of Zermelo–Fraenkel set theory, there will be *some* undecidable statements. Cohen's techniques enabled mathematicians to take specific mathematical statements and prove that those particular statements were undecidable. Cohen himself used the new technique to address the continuum problem, a famous question posed by Hilbert in 1900. Cohen showed that the question was not decidable.

Computability theory also began in the 1930s, and indeed, Gödel himself made major contributions to this field. From today's viewpoint, it is interesting to look back at the work on the concept of computability carried out two decades before any kind of real computer would be built, and fifty years before the arrival of today's desktop computers. In particular, Alan Turing, an English mathematician, proved an abstract theorem that established the theoretical possibility of a single computing machine that could be programmed to perform any computation you like. The American logician Stephen Cole Kleene proved another abstract theorem showing that the program for such a machine was essentially no different from the data it would run on.

All these areas of mathematical logic had in common the characteristic of being mathematical. By this I mean not just that the studies were carried out in a mathematical fashion, but that, by and large, their subject matter was mathematics itself. Thus, the enormous advances made in logic during this period were achieved at a price. Logic had begun with the ancient Greeks' attempts to analyze human reasoning in general, not just reasoning about mathematics. Boole's algebraic theory of logic brought the methods of mathematics to the study of reasoning, but the patterns of reasoning examined were arguably still general ones. However, the highly technical mathematical logic of the twentieth century was exclusively mathematical, both in the techniques used and the kinds of reasoning studied. In achieving mathematical perfection, logic had, to a large extent, broken away from its original goal of using mathematics to describe patterns of the human mind.

But while logicians were developing mathematical logic as a new branch of mathematics, the use of mathematics to describe patterns of the mind was being taken up once again. This time it was not the mathematicians who were doing the work, but a quite different group of scholars.

Patterns of language

To most people, it comes as a shock to discover that mathematics can be used to study the structure of language—the real, human languages they use in their everyday lives: English, Spanish, Japanese, and so on. Surely, ordinary language is not in the least mathematical, is it?

Take a look at A, B, and C below. In each case, without hesitation, decide whether you think that what you see is a genuine English sentence.

A. Biologists find *Spinelli morphenium* an interesting species to study.
B. Many mathematicians are fascinated by quadratic reciprocity.
C. Bananas pink because mathematics specify.

Almost certainly you decided, without having to give the matter any thought at all, that A and B are proper sentences but that C is not.

And yet A involves some words that you have never seen before. How can I be so sure? Because I made up the two words '*Spinelli*' and '*morphenium*'. So in fact, you happily classified as an English sentence a sequence of 'words', some of which are not really words at all!

In example B, all the words are genuine English words, and the sentence is in fact true. But unless you are a professional mathematician, you are unlikely to have ever come across the phrase 'quadratic reciprocity'. And yet again, you are quite happy to declare B to be a genuine sentence.

On the other hand, I am sure you had no hesitation in deciding that C is not a sentence, even though in this case you were familiar with all the words involved.

How did you perform this seemingly miraculous feat with so little effort? More precisely, just what is it that distinguishes examples A and B from example C?

It obviously has nothing to do with whether the sentences are true or not, or even with whether you understand what they are saying. Nor does it make any difference whether you know all the words in the sentence, or even whether they are genuine words or not. What counts is the overall structure of the sentence (or nonsentence, as the case may be). That is to say, the crucial feature is the way the words (or nonwords) are put together.

This structure is, of course, a highly abstract thing; you can't point to it the way you can point to the individual words or to the sentence. The best you can do is observe that examples A and B have the appropriate structure, but example C does not. And that is where mathematics comes in, for mathematics is the science of abstract structure.

The abstract structure of the English language that we rely upon, subconsciously and effortlessly, in order to speak and write to one another and to understand one another, is what is called the *syntactic structure* of English. A set of 'axioms' that describes that structure is called a *grammar* for the language. This way of looking at languages is relatively recent, and was inspired by the mathematical logic developed in the 1930s and 1940s.

Around the turn of the century, there was a shift from studies of the historical aspects of languages, of their roots and evolution (often referred to as historical linguistics or philology), to analyses of languages as communicative systems as they exist at any given point in time, regardless of their history. This kind of study is generally referred to as *synchronic linguistics*. Modern, mathematically based linguistics grew out of this development. The change from historical linguistics to the study of language as a system was due essentially to Mongin-Ferdinand de Saussure in Europe and to Frank Boas and Leonard Bloomfield in the United States.

Bloomfield, in particular, emphasized a scientific approach to linguistics. He was an active proponent of the philosophical position known as logical positivism, advocated by the philosopher Rudolf Carnap and the Vienna Circle. Inspired by the recent work in logic and the foundations of mathematics, in particular the Hilbert program, logical positivism attempted to reduce all meaningful statements to a combination of propositional logic and sense data (what you can see, hear, feel, or smell). Some linguists, in particular the American Zellig Harris, went even further than Bloomfield, suggesting that mathematical methods could be applied to the study of language.

The process of finding axioms that describe the syntactic structure of language was begun by the American linguist Noam Chomsky (see Figure 2.4), though the idea for such an approach had been proposed over a century earlier, by Wilhelm von Humboldt. "To write a grammar for a language," Chomsky suggested, "is to formulate a set of generalizations, i.e., a theory, to account for one's observations of the language."

FIGURE 2.4   Noam Chomsky of the Massachusetts Institute of Technology.

Chomsky's revolutionary new way of studying language was described in his book *Syntactic Structures*, published in 1957. Within a couple of years of its appearance, this short treatise—the text itself occupies
a mere 102 pages—transformed American linguistics, turning it from a
branch of anthropology into a mathematical science. (The effect in Europe was less dramatic.)

Let's take a look at a small fragment of a Chomsky-style grammar
for the English language. I should say at the outset that English is very
complex, and this example gives just seven of the many rules of English
grammar. But this is surely enough to indicate the mathematical nature
of the structure captured by a grammar.

$$DNP\ VP \rightarrow S$$
$$V\ DNP \rightarrow VP$$
$$P\ DNP \rightarrow PP$$
$$DET\ NP \rightarrow DNP$$
$$DNP\ PP \rightarrow DNP$$
$$A\ NP \rightarrow NP$$
$$N \rightarrow NP$$

In words, the first of these rules says that a definite noun phrase (*DNP*) followed by a verb phrase (*VP*) gives you a sentence (*S*); the second says that a verb (*V*) followed by a *DNP* gives you a *VP*; the third that a preposition (*P*) followed by a *DNP* gives you a prepositional phrase (*PP*); the next that a determiner (*DET*), such as the word 'the', followed by a noun phrase (*NP*) gives you a *DNP*. Given that *A* stands for adjective and *N* stands for noun, you can figure out the meaning of the last three rules for yourself.

In order to use the grammar to generate (or analyze) English sentences, all you need is a lexicon—a list of words—together with their linguistic categories. For example:

the $\rightarrow$ *DET*
to $\rightarrow$ *P*
runs $\rightarrow$ *V*
big $\rightarrow$ *A*
woman $\rightarrow$ *N*
car $\rightarrow$ *N*

Using the above grammar with this lexicon, it is possible to analyze the structure of the English sentence

The woman runs to the big car.

Such an analysis is most commonly represented in the form of what is called a *parse tree*, as shown in Figure 2.5. (Insofar as it resembles a real tree, a parse tree is upside down, with its 'root' at the top.)

At the top of the tree is the sentence. Each step down by one level from any point in the tree indicates the application of one rule of the grammar. For example, the very first step down, starting from the topmost point, represents an application of the rule

$$DNP\ VP \rightarrow S.$$

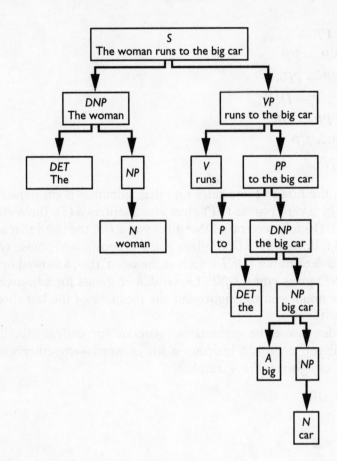

FIGURE **2.5** A parse tree.

The parse tree represents the abstract structure of the sentence. Any competent English speaker is able to recognize (generally subconsciously) such a structure. You may replace each of the words in this tree with other words, or even nonwords, and, provided your substitutions sound right for each grammatical category, the resulting sequence of words will sound like an English sentence. By providing axioms that determine all such parse trees, the formal grammar thus captures some of the abstract structure of English sentences.

The success of Chomsky's work does not mean that mathematics is able to capture everything there is to know about language. Mathematics never captures all there is to know about anything. Understanding gained through mathematics is just a part of a much larger whole. A human language such as English is a highly complex system, constantly changing and evolving. A grammar captures only one part of a much larger picture, but it is an important part. It is also one of the parts (there are others) that are best handled using mathematical techniques. English syntax is a complex, abstract structure, and mathematics is simply the most precise intellectual tool there is to describe abstract structures.

The fingerprint hidden in our words

Chomsky used algebra to capture some of the patterns of language that we all share. But there are other patterns that mathematicians can find in language. One of those patterns can be used to identify us from the words we write. Given a sufficiently long passage, a mathematician can determine who is most likely to have written it. What makes this possible is that the relative frequencies of the different words we typically use form a definite numerical profile that can identify us in much the same way, though with less accuracy, that we can be identified from our fingerprints.

One of the first applications of this means of identification was in 1962, when the American mathematicians Frederick Mosteller and David Wallace used the technique to resolve the issue of who wrote the various *Federalist* papers, a question of considerable interest to students of the origins of the United States Constitution.

The Federalist is a collection of eighty-five papers published from 1787 to 1788 by Alexander Hamilton, John Jay, and James Madison. Their goal was to persuade the people of New York State to ratify the new Constitution. Because none of the individual papers bore the name of its actual author, Constitutional historians were faced with a question: just who did write each paper? This question has considerable interest, since the papers provide insights into the men who formulated the Constitution and framed the future of the United States. For all but twelve of the papers, historical evidence provided the answer. It was generally agreed that fifty-one had been written by Hamilton, fourteen by Madison, and five by Jay. That left fifteen unaccounted for. Of these, twelve were in dispute between Hamilton and Madison, and three were believed to be joint works.

Mosteller and Wallace's strategy was to look for patterns in the writing—not the syntactic patterns studied by Chomsky and other linguists, but numerical patterns. As mentioned already, what made this approach possible is the fact that each individual has a distinctive style of writing, elements of which are susceptible to a statistical analysis. To determine its authorship, various numerical values associated with the disputed paper could be compared with those of other writings known to have come from the pens of the persons concerned.

One obvious number to examine is the average number of words an author uses in a sentence. Though this number can vary depending on the topic, when an author writes on a single topic, as is the case with the *Federalist* papers, the average sentence length remains remarkably constant from paper to paper.

In the *Federalist* case, however, this approach was far too crude. For the undisputed papers, Hamilton averaged 34.5 words per sentence and Madison averaged 34.6. It was impossible to tell the two authors apart simply by looking at sentence length.

A number of seemingly more subtle approaches, such as comparing the use of 'while' against 'whilst', also failed to lead to a definite conclusion.

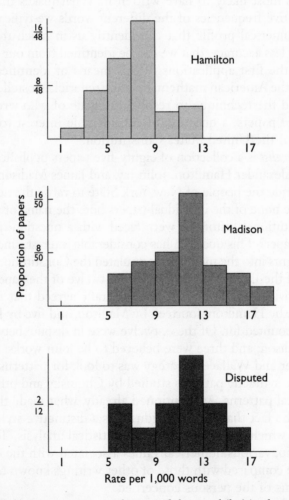

FIGURE **2.6** Frequency patterns in the use of the word 'by' in the writings of the authors of the *Federalist* papers.

What finally worked was to compare the relative frequency with which each author used each of thirty carefully chosen common words, among them 'by', 'to', 'this', 'there', 'enough', and 'according'. When the rates at which the three authors used each of these words were subjected to a computer analysis that looked for numerical patterns, the results were quite dramatic. Each author's writing exhibited a distinctive numerical 'fingerprint'.

For example, in his undisputed writings, Hamilton used 'on' and 'upon' almost equally, at the rate of about 3 times per 1,000 words. In contrast, Madison almost never used 'upon'. Hamilton used the word 'the' on average 91 times per 1,000 words and Madison 94 times per 1,000 words, which does not distinguish between them, but Jay's rate was 67 per 1,000, so the frequency of use of the word 'the' can help to distinguish between Jay and the other two. Figure 2.6 shows the different rates at which the authors used the word 'by'.

Taken on its own, the evidence of any one such word is suggestive, but hardly convincing. However, the detailed statistical analysis of the word rates carried out for all thirty words was far more reliable, and the possibility of error in the final conclusion was provably minuscule.

The conclusion was that, almost certainly, Madison had authored the disputed papers.

In an age when many people proclaim an inability to do mathematics, it is interesting to note that the work of linguists and statisticians shows that our very use of language involves mathematics, albeit subconsciously. As Chomsky demonstrated, the abstract patterns of grammatical sentences are mathematical—at least, they can best be described by mathematics—and Mosteller and Wallace's analysis of the *Federalist* papers shows us that when we write, we do so with a definite mathematical pattern of word frequencies every bit as unique to us as our fingerprints. Not only is mathematics the language of the universe, as Galileo observed, it may also be used to help us understand ourselves.

Mathematics in Motion

A world in motion

We live in a world that is in constant motion, much of it recognizably regular.

The sun rises each morning and sweeps a steady path across the daytime sky. As the seasons pass, the height of the sun's path above the horizon rises and falls, again in a regular manner.

A dislodged rock will roll down a hillside, and a stone thrown in the air will curve through the air before falling to the ground.

Moving air brushes against our faces, rain falls on our heads, the tides ebb and flow, the clear sky fills with drifting clouds, animals run and walk or fly or swim, plants spring from the ground, grow, and then die, diseases break out and spread through populations.

Motion is everywhere, and without it there would be no such thing as life. Still life exists only in the art gallery; it is not real life, for motion and change of one kind or another are the very essence of life.

Some motion appears chaotic, but much has order and regularity, exhibiting the kind of regular pattern that is, or at least ought to be, amenable to mathematical study. But the tools of mathematics are essentially static; numbers, points, lines, equations, and so forth do not in any way incorporate motion. So, in order to study motion, a way has to be found whereby these static tools can be brought to bear on patterns of change. It took some two thousand years of effort for humankind to achieve this feat. The biggest single step was the development of the calculus in the middle of the seventeenth century. This one mathematical advance marked a turning point in human history, having as dramatic and revolutionary an effect on our lives as the invention of the wheel or the printing press.

In essence, the calculus consists of a collection of methods to describe and handle patterns of infinity—the infinitely large and the infinitely small. For, as the ancient Greek philosopher Zeno indicated by a series of tantalizing paradoxes (as we shall see presently), the key to understanding the nature of motion and change is to find a way to tame infinity.

There is another paradox here: though infinity is not part of the world we live in, it seems that the human mind requires a mastery of infinity in order to analyze motion and change in the world. Perhaps, then, the methods of the calculus say as much about ourselves as they do about the physical world to which they can be applied with such effect. The patterns of motion and change we capture using the calculus certainly correspond to the motion and change we observe in the world, but, as patterns of infinity, their existence is inside our minds. They are patterns we human beings develop to help us comprehend our world.

A simple experiment illustrates a particularly striking number pattern that arises in motion. Take a long length of plastic guttering and fix it to form a descending ramp (see Figure 3.1). Place a ball at the top end and release it. Mark off the position of the ball after it has rolled for exactly 1 second. Now mark off the entire length of guttering into lengths equal to the first, and number the markings 1, 2, 3, . . . If you now release the ball from the top again and follow its descent, you will notice that after 1 second it has reached mark number 1, after 2 seconds it is

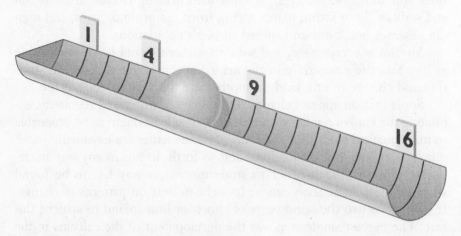

FIGURE **3.1** The rolling ball experiment.

at mark number 4, after 3 seconds it is at mark 9, and, if your ramp is long enough, after a further second's descent it is at mark 16.

The pattern here is obvious: after n seconds' descent, the ball is at the mark numbered n^2. And what is more, this is true regardless of the angle at which you incline the ramp.

Though it is simple to observe, a complete mathematical description of this pattern requires the full power of the differential and integral calculus, techniques to be explained in the remainder of this chapter.

The two men who invented calculus

Calculus was invented at about the same time by two mathematicians working independently of each other: Isaac Newton in England and Gottfried Wilhelm Leibniz in Germany (see Figure 3.2). Who were these men whose mathematical insight was to change forever the way men and women lived their lives?

Isaac Newton was born on Christmas Day, 1642, in the Lincolnshire village of Woolsthorpe. In 1661, following a fairly normal grammar school education, he entered Trinity College, Cambridge, where, largely through independent study, he acquired a mastery of astronomy and mathematics. In 1664 he was promoted to 'scholar', a status that provided him with four years of financial support toward a master's degree.

FIGURE 3.2 The fathers of the calculus: Sir Isaac Newton and Gottfried Leibniz.

It was on his return home to Woolsthorpe in 1665, when the university was forced to close because of the bubonic plague, that the twenty-three-year-old Newton embarked upon one of the most productive two years of original scientific thought the world has ever seen. The invention of the method of fluxions (his name for the differential calculus) and the inverse method of fluxions (the integral calculus) were just two of a flood of accomplishments in mathematics and physics that he made during the years 1665 and 1666.

In 1668, Newton completed his master's degree and was elected a Fellow of Trinity College, a lifetime position. A year later, when Isaac Barrow resigned his prestigious Lucasian Chair of Mathematics in order to become Chaplain to the King, Newton was appointed to the position.

An overwhelming fear of criticism kept Newton from publishing a great deal of his work, including the calculus, but in 1684 the astronomer Edmund Halley persuaded him to prepare for publication some of his work on the laws of motion and gravitation. The eventual appearance, in 1687, of *Philosophiae Naturalis Principia Mathematica* was to change physical science for all time, and established Newton's reputation as one of the most brilliant scientists the world had—and has—ever seen.

In 1696, Newton resigned his Cambridge chair to become Warden of the Royal Mint. It was while he was in charge of the British coinage that he published, in 1704, his book *Opticks*, a mammoth work outlining the optical theories he had been working on during his Cambridge days. In an appendix to this book he gave a brief account of the method of fluxions that he had developed forty years previously. It was the first time he had published any of this work. A more thorough account, *De Analysi*, had been in private circulation among the British mathematical community from the early 1670s onward, but was not published until 1711. A complete account of the calculus written by Newton did not appear until 1736, nine years after his death.

Just prior to the appearance of *Opticks*, Newton was elected President of the Royal Society, the ultimate scientific accolade in Great Britain, and in 1705, Queen Anne bestowed on him a knighthood, the ultimate royal tribute. The once shy, frail boy from a small Lincolnshire village was to spend the remaining years of his life regarded as little less than a national treasure.

Sir Isaac Newton died in 1727 at the age of eighty-four, and was buried in Westminster Abbey. His epitaph in the Abbey reads: "Mortals, congratulate yourselves that so great a man has lived for the honor of the human race."

The other inventor of calculus, Gottfried Wilhelm Leibniz, was born in Leipzig in 1646. He was a child prodigy, taking advantage of the sizable scholarly library of his father, a professor of philosophy. By the time he was fifteen years old, the young Leibniz was ready to enter the University of Leipzig. Five years later, he had completed his doctorate, and was set to embark on an academic career, when he decided to leave university life and enter government service.

In 1672, Leibniz became a high-level diplomat in Paris, from where he made a number of trips to Holland and Britain. These visits brought him into contact with many of the leading academics of the day, among them the Dutch scientist Christian Huygens, who inspired the young German diplomat to take up once more his studies in mathematics. It proved to be a fortuitous meeting, for, by 1676, Leibniz had progressed from being a virtual novice in mathematics to having discovered for himself the fundamental principles of calculus.

Or had he? When Leibniz first published his findings in 1684, in a paper in the journal *Acta Eruditorum*, for which he was the editor, many of the British mathematicians of the time cried foul, accusing Leibniz of taking his ideas from Newton. Certainly, on a visit to the Royal Society in London in 1673, Leibniz had seen some of Newton's unpublished work, and in 1676, in response to a request for further information about his discoveries, Newton had written two letters to his German counterpart, providing some of the details.

Though the two men themselves largely stayed out of the debate, the argument between British and German mathematicians over who had invented calculus grew heated at times. Certainly, Newton's work had been carried out before Leibniz's, but the Englishman had not published any of it. In contrast, not only had Leibniz published his work promptly, but his more geometric approach led to a treatment that is in many ways more natural, and which quickly caught on in Europe. Indeed, to this day, Leibniz's geometric approach to differentiation (page 116) is the one generally adopted in calculus classes the world over, and Leibniz's notation (dy/dx, as we shall see presently) for the derivative (page 116) is in widespread use, whereas Newton's approach in terms of physical motion and his notation are rarely used outside of physics.

Today, the general opinion is that, although Leibniz clearly obtained some of his ideas from reading part of Newton's work, the German's contribution was undoubtedly significant enough to grant both men the title of 'father of the calculus'.

Like Newton, Leibniz was not content to spend his entire life working in mathematics. He studied philosophy, he developed a theory of formal logic, a forerunner of today's symbolic logic, and he became an expert in the Sanskrit language and the culture of China. In 1700, he was a major force in the creation of the Berlin Academy, of which he was president until his death in 1716.

In contrast to Newton, who was given a state funeral in Westminster Abbey, Germany's creator of the calculus was buried in quiet obscurity.

So much the for the men who created calculus. What exactly is it? The story begins, as so often in mathematics, back in ancient Greece.

The paradox of motion

Calculus applies to continuous, rather than discrete, motion. But on first analysis, the very idea of continuous motion seems to be paradoxical. Consider: At a particular instant in time, any object must be at a particular location, in a certain position in space. At that instant, the object is indistinguishable from a similar object at rest. But this will be true of any instant in time, so how can the object move? Surely, if the object is at rest at every instant, then it is always at rest (see Figure 3.3).

This particular paradox of motion was first put forward by the Greek philosopher Zeno, probably as an argument against the numerically based mathematical studies of the Pythagoreans. Zeno, who lived about 450 B.C., was a student of Parmenides, the founder of the Eleatic school of philosophy, which flourished for a while in Elea, in Magna Graecia. Expressed originally in terms of an arrow in flight, Zeno's puzzle is a gen-

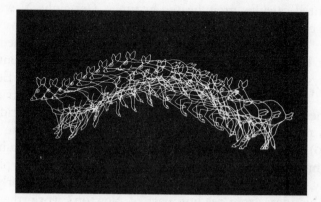

FIGURE **3.3** The paradox of motion. At any one instant, an object must be at rest, an idea captured by this illustration of a leaping deer. Since this is true for all instants, surely the object will always be at rest, so how can motion arise? The Greek philosopher Zeno posed this paradox as a challenge to the belief that time consists of a succession of discrete instants.

uine paradox if one regards space as atomic, consisting of a multiplicity of adjacent points, and time as consisting of a succession of discrete instants.

Another of Zeno's puzzles creates a paradox for those who believe that space and time are not atomic, but infinitely divisible. This is the paradox of Achilles and the tortoise, perhaps the best known of Zeno's arguments. Achilles is to race a tortoise over a course of 100 meters. Since Achilles can run ten times faster than the tortoise, the tortoise is given a 10-meter head start. The race starts, and Achilles sets off in pursuit of the tortoise. In the time it takes Achilles to cover the 10 meters to the point where the tortoise started, the tortoise has covered exactly 1 meter, and so is 1 meter ahead. By the time Achilles has covered that extra meter, the tortoise is a tenth of a meter in the lead. When Achilles gets to that point, the tortoise is a hundredth of a meter ahead. And so on, ad infinitum. Thus, the argument goes, the tortoise remains forever in the lead, albeit by smaller and smaller margins; Achilles never overtakes his opponent to win the race.

The purpose of these paradoxes was certainly not to argue that an arrow cannot move, or that Achilles can never overtake the tortoise. Both of these are undeniable, empirical facts. Rather, Zeno's puzzles presented challenges to the attempts of the day to provide analytic explanations of space, time, and motion—challenges that the Greeks themselves were not able to meet. Indeed, truly satisfactory resolutions to these paradoxes were not found until the end of the nineteenth century, when mathematicians finally came to grips with the mathematical infinite.

Taming infinity

The key to the eventual development of a mathematical treatment of motion and change was finding a way to handle infinity. And that meant finding ways to describe and manipulate the various patterns that involve infinity.

For example, Zeno's paradox of Achilles and the tortoise can be disposed of once you have a way of dealing with the pattern involved. The amounts by which the tortoise is ahead of Achilles at each stage of the race are (in meters)

$$10, 1, \frac{1}{10}, \frac{1}{100}, \frac{1}{1,000}, \ldots$$

Thus the paradox hinges on what we make of the infinite sum

$$10 + 1 + \frac{1}{10} + \frac{1}{100} + \frac{1}{1,000} + \ldots$$

where the ellipsis (those three dots) signifies that this sum goes on for-ever, following the pattern indicated.

There is no hope of actually adding together all the infinitely many terms in this sum. Indeed, I can't even write it out in full, so my use of the word 'sum' might be misleading; it is not a sum in the normal sense of the word. In fact, to avoid such confusion, mathematicians re-fer to such infinite sums as *infinite series*. This is one of a number of instances in which mathematicians have taken an everyday word and given it a technical meaning, often only slightly related to its everyday use.

By shifting our attention from the individual terms in the series to the overall pattern, it is easy to find the value of the series. Let S denote the unknown value:

$$S = 10 + 1 + \frac{1}{10} + \frac{1}{100} + \frac{1}{1,000} + \ldots$$

The pattern in this series is that each successive term is one-tenth the previous term. So, if you multiply the entire series through by 10, you obtain the same series again, apart from the first term:

$$10S = 100 + 10 + 1 + \frac{1}{10} + \frac{1}{100} + \frac{1}{1,000} + \ldots$$

If you now subtract the first identity from this second one, all the terms on the right-hand side cancel out in pairs, apart from the initial 100 in the second series:

$$10S - S = 100.$$

Now you have an ordinary, finite equation, which can be solved in the usual way:

$$9S = 100,$$

so

$$S = \frac{100}{9} = 11\tfrac{1}{9}.$$

In other words, Achilles draws level with the tortoise when he has cov-ered exactly $11\tfrac{1}{9}$ meters.

The crucial point is that an infinite series can have a finite value; Zeno's puzzle is paradoxical only if you think that an infinite series must have an infinite value.

Notice that the key to finding the value of the series was to shift attention from the process of adding the individual terms to the identification and manipulation of the overall pattern. In a nutshell, this is the key to handling the infinite in mathematics.

Infinity bites back

As you might have realized, it is not entirely clear that one may justifiably multiply through an infinite series by a fixed number, as I did above, or that one may then subtract one series from another, term by term, as I also did. Infinite patterns are notoriously slippery customers, and it is easy to go wrong in manipulating them. Take a look at the following infinite series, for instance:

$$S = 1 - 1 + 1 - 1 + 1 - 1 + \ldots$$

If you multiply through the series by -1, you obtain the same series with all terms shifted one along:

$$S = 1 - 1 + 1 - 1 + 1 - 1 + \ldots$$
$$-S = - 1 + 1 - 1 + 1 - 1 + \ldots$$

If you then subtract the second series from the first, all the terms on the right cancel out, apart from the first term of the first series, to leave

$$2S = 1.$$

The conclusion is that $S = \frac{1}{2}$.

All well and good, you might think. But now suppose you take the original series and pair off the terms like this:

$$S = (1 - 1) + (1 - 1) + (1 - 1) + \ldots$$

Again, this seems a perfectly reasonable thing to do to the overall pattern; though there are infinitely many terms in the series, I have described the pattern whereby the bracketing is done. But this time, each bracketed pair works out to be 0, so now the conclusion is that

$$S = 0 + 0 + 0 + \ldots$$

which means that $S = 0$.

Or you could apply brackets according to the following pattern:

$$S = 1 + (-1 + 1) + (-1 + 1) + (-1 + 1) + \ldots$$

This time you obtain the value $S = 1$.

The original series for S was given by a perfectly understandable pattern. I manipulated it in three different ways, using three different patterns of manipulation, and arrived at three different answers: $S = \frac{1}{2}$, 0, 1. Which answer is correct?

In fact, there is no correct answer. The pattern of this series cannot be handled in a mathematical fashion: this particular infinite series simply does not have a value. On the other hand, the series that arises from Achilles and the tortoise does have a value, and indeed, the manipulation I carried out in that case is permissible. Sorting out the distinction between series that can be manipulated and those that cannot, and developing a sound theory of how to handle infinite series, took hundreds of years of effort, and was not completed until late in the nineteenth century.

A particularly elegant illustration of the manner in which the value of an infinite series can be determined by manipulating the pattern of the series is provided by so-called *geometric series*. These are series of the form

$$S = a + ar + ar^2 + ar^3 + \ldots$$

where each successive term is obtained by multiplying the previous term by some fixed amount r. Geometric series arise frequently in everyday life, for example, in radioactive decay and in the computation of the interest you must pay on a bank loan or a mortgage. The series that arose in the paradox of Achilles and the tortoise was also a geometric series. (The fixed ratio r was $\frac{1}{10}$.) In fact, the method I used to find the value of that series works for any geometric series. To obtain the value of S, you multiply through the series by the common ratio r, giving the new series

$$Sr = ar + ar^2 + ar^3 + ar^4 + \ldots$$

and subtract the new series from the first. All the terms cancel out in pairs, apart from the initial term a in the first series, leaving the equation

$$S - Sr = a.$$

Solving this equation for S, you obtain $S = a/(1 - r)$. The only remaining question is whether the various manipulations just described are valid or not. A more detailed examination of the pattern indicates that these manipulations are permissible in the case where r is less than 1 (in the case of a negative r, it must be greater than -1), but are not valid for other values of r.

So, for example, the series

$$S = 1 + \frac{1}{2} + \frac{1}{4} + \frac{1}{8} + \frac{1}{16} + \ldots + \frac{1}{2^n} + \ldots$$

has initial term $a = 1$ and ratio $r = \frac{1}{2}$, so its value is

$$S = \frac{1}{1 - \frac{1}{2}} = \frac{1}{\frac{1}{2}} = 2.$$

Obviously, one consequence of the ratio r being less than 1 (and more than -1 in the case where r is negative) is that the terms in the series get smaller as it progresses. Could this be the crucial factor that enables you to find a finite value for an infinite series?

On the face of it, this hypothesis seems reasonable; if the terms get progressively smaller, their effect on the sum becomes increasingly insignificant. If this is indeed the case, then it will follow that the following elegant series has a finite value:

$$S = 1 + \frac{1}{2} + \frac{1}{3} + \frac{1}{4} + \ldots + \frac{1}{n} + \ldots$$

Because of its connection to certain patterns on the musical scale, this series is known as the *harmonic series*.

If you add together the first thousand terms, you obtain the value 7.485 (to three places of decimals); the first million terms add together to give 14.357 (to three places); the first billion give approximately 21, and the first trillion about 28. But what is the value of the entire, infinite sum?

The answer is that there is no value, a result first discovered by Nicolae Oresme in the fourteenth century. Thus, the fact that the terms of an infinite series get progressively smaller is not in itself enough to guarantee that the series has a finite value.

How do you set about proving that the harmonic series does not have a finite value? Certainly not by adding together more and more terms. Suppose you were to start to write out the series term by term on a ribbon, allowing one centimeter for each term (a gross underestimation, since you would need to write down more and more digits the further along in the series you went). You would need some 10^{43} centimeters of ribbon to write down enough terms to sum to a value that exceeds 100. But 10^{43} centimeters is about 10^{25} light-years, which exceeds the known size of the entire universe (for which 10^{12} light-years is one current estimate).

The way to show that the harmonic series has an infinite value is to work with the pattern, of course. Start off by observing that the third and fourth terms are both at least $\frac{1}{4}$, so their sum is at least $2 \times \frac{1}{4} = \frac{1}{2}$. Now notice that the next four terms, namely, $\frac{1}{5}, \frac{1}{6}, \frac{1}{7}, \frac{1}{8}$, are all at least $\frac{1}{8}$, so their sum is at least $4 \times \frac{1}{8} = \frac{1}{2}$. Likewise, the next sixteen terms, from $\frac{1}{9}$ to $\frac{1}{32}$, are all at least $\frac{1}{32}$, so they also add up to at least $16 \times \frac{1}{32} = \frac{1}{2}$. By taking increasingly longer groups of terms, according to the pattern 2 terms, 4 terms, 8 terms, 16 terms, 32 terms, and so on, you keep on getting sums that in each case are at least $\frac{1}{2}$. This procedure will lead to infinitely many repetitions of $\frac{1}{2}$, and adding together infinitely many $\frac{1}{2}$s will surely produce an infinite result. But the value of the harmonic series, if there is indeed a value, will be at least as big as this infinite sum of $\frac{1}{2}$s. Hence the harmonic series cannot have a finite value.

During the seventeenth and eighteenth centuries, mathematicians became ever more skilled at manipulating infinite series. For instance, the Scotsman James Gregory found the following result in 1671:

$$\frac{\pi}{4} = \frac{1}{1} - \frac{1}{3} + \frac{1}{5} - \frac{1}{7} + \frac{1}{9} - \cdots$$

You might well wonder how this infinite series gives you an answer that involves the mathematical constant π, the ratio of the circumference of any circle to its diameter.

In 1736, Euler discovered another infinite series whose value involved π:

$$\frac{\pi^2}{6} = \frac{1}{1^2} + \frac{1}{2^2} + \frac{1}{3^2} + \frac{1}{4^2} + \frac{1}{5^2} + \cdots$$

In fact, Euler went on to write a whole book on infinite series, *Introductio in Analysin Infinitorum*, which was published in 1748.

By concentrating on patterns rather than arithmetic, mathematicians were thus able to handle the infinite. The most significant consequence of the study of infinite patterns took place in the second half of the seventeenth century, when Newton and Leibniz developed the differential calculus. Their achievement is undoubtedly one of the greatest mathematical feats of all time, and it transformed human life forever. Without the differential calculus, modern technology simply would not exist; there would be no electricity, no telephones, no automobiles, and no heart-bypass surgery. The sciences that led to these—and most other—technological developments depend on the calculus in a fundamental way.

Functions provide the key

The differential calculus provides a means of describing and analyzing motion and change—not just any motion or change, but motion or change that shows a pattern. To use the differential calculus, you must be presented with a pattern that describes the motion or change you are interested in. This is because, in concrete terms, the differential calculus is a collection of techniques for the manipulation of patterns. (The word *calculus* is a Latin word that means 'pebble'—recall that early counting systems involved the physical manipulation of pebbles.)

The basic operation of the differential calculus is a process known as *differentiation*. The aim of differentiation is to obtain the rate of change of some changing quantity. In order to do this, the value, or position, or path of that quantity has to be given by means of an appropriate formula. Differentiation then acts upon that formula to produce another formula that gives the rate of change. Thus, differentiation is a process for turning formulas into other formulas.

For example, imagine a car traveling along a road. Suppose that the distance it travels along the road, say x, varies with time, t, according to the formula

$$x = 5t^2 + 3t.$$

Then, according to the differential calculus, the car's speed s (i.e., the rate of change of position) at any time t is given by the formula

$$s = 10t + 3.$$

The formula $10t + 3$ is the result of differentiating the formula $5t^2 + 3t$. (You will see shortly just how differentiation works in this case.)

Notice that the speed of the car is not a constant in this example; the speed varies with time, just as the distance does. Indeed, the process of differentiation may be applied a second time to obtain the acceleration (i.e., the rate of change of the speed). Differentiating the formula $10t + 3$ produces the acceleration

$$a = 10,$$

which in this case is a constant.

The fundamental mathematical objects to which the process of differentiation is applied are called *functions*. Without the notion of a function, there can be no calculus. Just as arithmetical addition is an operation that is performed on numbers, so differentiation is an operation that is performed on functions.

But what exactly is a function? The simplest answer is that, in mathematics, a function is a rule that, given one number, allows you to calculate another. (Strictly speaking, this is a special case, but it is adequate for understanding how the calculus works.)

For example, a polynomial formula such as

$$y = 5x^3 - 10x^2 + 6x + 1$$

determines a function. Given any value for x, the formula tells you how to compute a corresponding value for y. For instance, given the value $x = 2$, you may compute

$$y = 5 \times 2^3 - 10 \times 2^2 + 6 \times 2 + 1 = 40 - 40 + 12 + 1 = 13.$$

Other examples are the *trigonometric functions*, $y = \sin x$, $y = \cos x$, $y = \tan x$. For these functions, there is no simple way to compute the value of y, as we could in the case of a polynomial. Their familiar definitions are given in terms of the ratios of the various sides of right-angled triangles, but those definitions apply only when the given x is an angle less than a right angle. The mathematician defines the tangent function in terms of the sine and cosine functions as

$$\tan x = \frac{\sin x}{\cos x},$$

and defines the sine and cosine functions by means of infinite series:

$$\sin x = x - \frac{x^3}{3!} + \frac{x^5}{5!} - \frac{x^7}{7!} + \ldots$$
$$\cos x = 1 - \frac{x^2}{2!} + \frac{x^4}{4!} - \frac{x^6}{6!} + \ldots$$

To understand these formulas, you need to know that, for any natural number n, $n!$ (read "n-factorial") is equal to the product of all numbers from 1 to n inclusive: for example, $3! = 1 \times 2 \times 3 = 6$. The infinite series for $\sin x$ and $\cos x$ always give a finite value, and may be manipulated more or less like finite polynomials. The series give the usual values when x is an angle of a right-angled triangle; but they also give a value for *any* real number x.

Still another example of a function is the exponential function

$$e^x = 1 + \frac{x^1}{1!} + \frac{x^2}{2!} + \frac{x^3}{3!} + \frac{x^4}{4!} + \ldots$$

Again, this infinite series always gives a finite value, and may be manipulated like a finite polynomial. Using $x = 1$, you get

$$e = e^1 = 1 + \frac{1}{1!} + \frac{1}{2!} + \frac{1}{3!} + \frac{1}{4!} + \ldots$$

The mathematical constant e that is the value of this infinite series is an irrational number. Its decimal expansion begins with 2.71828.

The exponential function e^x has an important inverse function—that is to say, a function that exactly reverses the effect of e^x. It is the natural logarithm of x, $\ln x$. If you start with a number a and apply the function e^x to get the number $b = e^a$, then, when you apply the function $\ln x$ to b, you get a again: $a = \ln b$.

How to compute slopes

Algebraic formulas such as polynomials or the infinite series for the trigonometric or exponential functions provide a very precise, and extremely useful, way of describing a certain kind of abstract pattern. The pattern in these cases is a pattern of association between pairs of numbers: the independent variable or argument, x, that you start with and the dependent variable or value, y, that results. In many cases, this pattern can be illustrated by means of a graph, as in Figure 3.4. The graph of a function shows at a glance how the variable y is related to the variable x.

For example, in the case of the sine function, as x increases from 0, y also increases, until somewhere near $x = 1.5$ (the exact point is $x = \frac{\pi}{2}$), y starts to decrease; y becomes negative around $x = 3.1$ (precisely, when $x = \pi$), continues to decrease until around $x = 4.7$ (precisely, $x = \frac{3\pi}{2}$), then starts to increase again.

The task facing Newton and Leibniz was this: How do you find the rate of change of a function such as $\sin x$—that is, how do you find the rate of change of y with respect to x? In terms of the graph, finding the rate of change is the same as finding the slope of the curve—how steep is it? The difficulty is that the slope is not constant: at some points the curve is climbing fairly steeply (a large, positive slope), at other points it is almost horizontal (a slope close to zero), and at still other points it is falling fairly steeply (a large, negative slope).

In summary, just as the value of y depends on the value of x, so too does the slope at any point depend on the value of x. In other words, the slope of a function is itself a function—a second function. The question now is, given a formula for a function—that is to say, a formula that

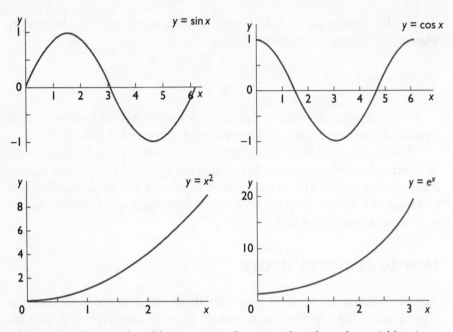

FIGURE **3.4** The graphs of four common functions show how the variable y is related to the variable x.

describes the pattern relating x to y—can you find a formula that describes the pattern relating x to the slope?

The method that both Newton and Leibniz came up with is, in essence, as follows. For simplicity, let us consider the function $y = x^2$, whose graph is shown in Figure 3.5. As x increases, not only does y increase, but the slope also increases. That is, as x increases, not only does the curve climb higher, it also gets steeper. Given any value of x, the height of the curve for that value of x is given by computing x^2. But what do you do to x in order to compute the slope for that value of x?

Here's the idea. Look at a second point a short distance h to the right of x. Referring to Figure 3.5, the height of point P on the curve is x^2, and the height of point Q is $(x + h)^2$. The curve bends upward as you go from P to Q. But if h is fairly small (as shown), the difference between the curve and a straight line joining P to Q is also small. So the slope of the curve at P will be close in value to the slope of this straight line.

The point of this move is that it is easy to compute the slope of a straight line: you just divide the increase in height by the increase in the horizontal direction. In this case, the increase in height is

$$(x + h)^2 - x^2,$$

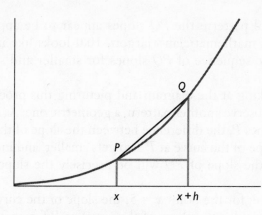

FIGURE **3.5** Computing the slope of the function $y = x^2$.

and the increase in the horizontal direction is h, so the slope of the straight line from P to Q is

$$\frac{(x + h)^2 - x^2}{h}.$$

Using elementary algebra, the numerator in this fraction simplifies as follows:

$$(x + h)^2 - x^2 = x^2 + 2xh + h^2 - x^2 = 2xh + h^2.$$

So the slope of the straight line PQ is

$$\frac{2xh + h^2}{h}.$$

Canceling the h from this fraction leaves $2x + h$.

This is a formula for the slope of the straight line from P to Q. But what about the slope of the curve $y = x^2$ at the point P, which is what you started out trying to compute? This is where Newton and Leibniz made their brilliant and decisive move. They argued as follows: Replace the static situation with a dynamic one, and think about what would happen if the distance h that separates the two points P and Q along the x-direction got smaller and smaller.

As h gets smaller, point Q moves closer and closer to point P. Remember that, for each value of h, the formula $2x + h$ gives the corresponding value for the slope of the straight line PQ. For instance, if you take $x = 5$, and let h successively assume each of the values 0.1, 0.01, 0.001, 0.0001, and so on, then the corresponding PQ slopes are 10.1, 10.01, 10.001, 10.0001, ... And there at once you see an obvious numerical pattern: the PQ slopes appear to be approaching the value 10.0. (In mathematician's jargon, 10.0 looks like it is the *limiting value* of the sequence of PQ slopes for smaller and smaller values of h.)

But by looking at the diagram and picturing this process geometrically, you can observe another pattern, a geometric one: As h gets smaller, and Q approaches P, the difference between the slope of the straight line PQ and the slope of the curve at P also gets smaller, and indeed, the limiting value of the slope of PQ will be precisely the slope of the curve at P.

For instance, for the point $x = 5$, the slope of the curve at P will be 10.0. More generally, the slope of the curve at P for an arbitrary point x will be $2x$. In other words, the slope of the curve at x is given by the formula $2x$ (which is the limiting value of the expression $2x + h$ as h approaches 0).

Ghosts of departed quantities

For the record, Newton's approach was not exactly as described above. His main concern was physics; in particular, he was interested in planetary motion. Instead of thinking of a variable y varying with a variable x in a geometric fashion as represented on a graph, Newton thought of a distance r (for radius) varying with time t—say, $r = t^2$. He referred to the given function as the *fluent* and the slope function as the *fluxion*. (So, if the fluent is $r = t^2$, the fluxion will be $2t$.) Obviously, the fluxion in this case would be some form of speed or velocity (i.e., a rate of change of distance). For the small increment that I earlier denoted by h, Newton used the symbol o, to suggest a quantity that was close to, but not quite equal to, 0.

Leibniz, on the other hand, approached the issue as a geometric problem of finding the slopes of curves, which is the approach I adopted above. He used the notation dx in place of my h, and dy to denote the corresponding small difference in y values (the difference in height between P and Q). He denoted the slope function by dy/dx, a notation obviously suggestive of a ratio of two small increments. (The notation dx is normally read "dee-ex," dy is read "dee-wye," and dy/dx is read "dee-wye by dee-ex.")

For both men, however, the important starting point was a functional relation linking the two quantities:

$r =$ some formula involving t

in Newton's case or

$y =$ some formula involving x

for Leibniz. In modern terminology, we say that r is a function of t, or that y is a function of x, and use notations such as $r = f(t)$ or $r = g(t)$, and analogously in the x, y version.

Motivation and notation aside, the crucial step taken by both Newton and Leibniz was to shift attention from the essentially static situation of the slope at a particular point P to the dynamic process of the successive approximation of that slope by slopes of straight lines starting from P. It was by observing numerical and geometric patterns in this *process* of approximation that Newton and Leibniz were able to arrive at the right answer.

Moreover, their approach works for a great many functions, not just for the simple example considered above. For example, if you start with the function x^3, you get the slope function $3x^2$, and, more generally, if you start with the function x^n, where n is any natural number, the slope function works out to be nx^{n-1}. With this, you have another easily recognizable, if somewhat unfamiliar, pattern, the pattern that takes x^n to nx^{n-1} for any value of n. This, as we shall see, is a pattern of differentiation.

It should be stressed that what Newton and Leibniz were doing was not at all the same as setting the value of h equal to 0. True enough, in the case of the very simple example above, where the function is x^2, if you simply set $h = 0$ in the slope formula $2x + h$, then you obtain $2x$, which is the right answer. But if $h = 0$, then the points Q and P are one and the same, so there is no straight line PQ. Remember that, although a factor of h was canceled to obtain a simplified expression for the slope of PQ, this slope is the ratio of the two quantities $2xh + h^2$ and h, and if you set $h = 0$, then this ratio reduces to the division of 0 by 0, which is meaningless.

This point was a source of considerable misunderstanding and confusion, both at the time Newton and Leibniz were doing their work and for many subsequent generations. To modern mathematicians, used to regarding mathematics as the science of patterns, the idea of looking for numerical and geometric patterns in a process of successive approximation is not at all strange, but back in the seventeenth century, even Newton and Leibniz could not formulate their ideas with sufficient precision to silence their many critics. The most notable of those critics was the English philosopher Bishop Berkeley, who, in 1734, published a stinging critique of the calculus.

Leibniz struggled to make himself clear by describing his *dx* and *dy* as "infinitely small" and "indefinitely small" quantities. When he failed to come up with a sound argument to support his manipulations of these entities, he wrote,

> [You may] think that such things are utterly impossible; it will be sufficient simply to make use of them as a tool that has advantages for the purpose of calculation.

While not going so far as to mention the 'infinitely small', Newton referred to his fluxion as the "ultimate ratio of evanescent increments," to which Berkeley retorted in his 1734 critique,

> And what are these fluxions? The velocities of evanescent increments. And what are these same evanescent increments? They are neither finite quantities, nor quantities infinitely small, nor yet nothing. May we not call them ghosts of departed quantities?

If you think of what Newton and Leibniz were doing in a static fashion, with *h* a small but fixed quantity, then Berkeley's objections are entirely correct. But if you regard *h* as a variable, and concentrate not on the given function, but rather on the process of approximation that arises when *h* approaches 0, then Berkeley's argument can no longer be sustained.

To construct a reliable defense against Berkeley's objections, you have to work out a rigorous mathematical theory of approximation processes, which neither Newton nor Leibniz was able to do. Indeed, it was not until 1821 that the Frenchman Augustin-Louis Cauchy developed the key idea of a *limit* (see presently), and it was a few years later still that the German Karl Weierstrass provided a formal definition of this notion. Only then was the calculus placed on a sound footing, almost two hundred years after its invention.

Why did the formulation of a rigorous theory take so long, and more intriguingly, how was it possible to develop such a powerful, and reliable, tool as the calculus without being able to provide a logical explanation of why it worked?

Chasing sound intuitions

Newton's and Leibniz's methods worked because the intuitions that drove the two men were sound: they knew that they were working with a dynamic process of successive approximation. Indeed, in his book *Principia*, Newton came very close to achieving the correct formulation with this explanation:

Ultimate ratios in which quantities vanish are not, strictly speaking, ratios of ultimate quantities, but limits to which the ratios of these quantities, decreasing without limit, approach.

In other words, to find the slope function for, say, x^2, it is permissible to determine what happens to the ratio $(2xh + h^2)/h$ as h approaches 0, but you may not set $h = 0$. (Note that cancellation of h to give the expression $2x + h$ for the ratio is permissible only if $h \neq 0$.)

But neither Newton nor Leibniz, nor anyone else until Cauchy and Weierstrass, was able to capture the notion of a limit in a precise mathematical fashion. And the reason is that they were not able to 'step back' quite far enough to discern the appropriate pattern in a static way. Remember, the patterns captured by mathematics are static things, even if they are patterns of motion. Thus, if Newton were thinking about, say, the motion of a planet whose position varies according to the square of the time, he would capture this dynamic situation by means of the static formula x^2—'static' because it simply represents a relationship between pairs of numbers. The dynamic motion is captured by a static function.

The key to putting the differential calculus on a rigorous footing was to observe that the same idea can be applied to the process of approximation to the slope. The dynamic process of obtaining closer and closer approximations to the slope as the increment h approaches zero can also be captured in a static fashion, as a function of h. Here is how Weierstrass did just that.

Suppose you have some function $f(h)$; in the example we have been looking at, $f(h)$ would be the quotient $(2xh + h^2)/h$ (where x is regarded as fixed, h as the variable). Then, to say that a number L ($2x$ in our example) is the limit of the function $f(h)$ as h approaches 0 means, precisely,

For any $\varepsilon > 0$, there exists a $\delta > 0$ so that, if $0 < |h| < \delta$, then $|f(h) - L| < \varepsilon$.

Unless you have seen this statement before, you are unlikely to be able to fathom its meaning. It did, after all, take some two hundred years for mathematicians to arrive at this definition. The point to notice, though, is that it says nothing about any kind of (dynamic) process; it simply refers to the existence of numbers δ having a certain property. In this respect, it is just the same as the original step, which Newton performed all the time, of capturing motion by means of a formula. By denoting time by a (static) variable t, Newton could capture motion by means of a formula involving t. By likewise treating h as a variable, Weierstrass was able to capture the notion of a limit (of a sequence of approximations) by means of a formal definition involving h. Newton captured the t-pattern, Weierstrass captured the h-pattern.

Incidentally, though Cauchy developed an extensive theory of limits, adequate for the calculus, he was still working with the notion of a dynamic approximation process. Thus, he put the calculus on a firm footing only in the sense that he reduced the problem to that of providing a precise definition of limit. That final, key step was performed by Weierstrass. But why wasn't Newton or Leibniz, or even Cauchy, able to do this? After all, each of these great mathematicians was well accustomed to the use of variables to capture motion and the use of formulas to capture patterns of motion. At issue, almost certainly, is the level of process that the human mind can cope with as an entity in itself. At the time of Newton and Leibniz, regarding a function as an entity, rather than as a process of change or motion, was already a significant cognitive accomplishment. To subsequently regard the process of successively approximating the slope of that function as yet another entity in its own right was simply too much. Only with the passage of time, and growing familiarity with the techniques of the calculus, could anyone make this second conceptual leap. Great mathematicians can perform amazing feats, but they are still only human. Cognitive advances take time, often many generations.

Since their intuitions concerning approximation (or limit) processes were so good, Newton and Leibniz were able to develop their differential calculus into a reliable and extremely powerful tool. To do this, they regarded functions as mathematical objects to be studied and manipulated, not just as recipes for computations. They were guided by the various patterns that arose in computing successive approximations to slopes (and other changing quantities) associated with those functions; but they were not able to step back and regard those patterns of approximations as being themselves objects for mathematical study.

The differential calculus

As we have seen, the process of going from the formula for a curve to a formula for the slope of that curve is known as *differentiation*. (The name reflects the idea of taking small *differences* in the x and y directions and computing the slopes of the resulting straight lines.) The slope function is called the *derivative* of the initial function (from which it is *derived*).

For the simple example we have been looking at, the function $2x$ is the derivative of the function x^2. Similarly, the derivative of the function x^3 is $3x^2$, and in general, the derivative of the function x^n for any natural number n is nx^{n-1}.

The power of Newton's and Leibniz's invention was that the number of functions that could be differentiated was greatly enlarged by the development of a *calculus*, a series of rules for differentiating complicated functions. The development of this calculus also accounts for the method's enormous success in different applications, despite its dependence on methods of reasoning that were not fully understood. People knew *what* to do, even if they did not know why it worked. Many of the students in today's calculus classes have a similar experience.

The rules of the calculus are most conveniently described using modern terminology, whereby arbitrary functions of x are denoted by expressions such as $f(x)$ or $g(x)$, and their derivatives (which are also functions of x) are denoted by $f'(x)$ and $g'(x)$, respectively. So, for example, if $f(x)$ is used to denote the function x^5, then $f'(x) = 5x^{5-1} = 5x^4$.

One of the rules of the calculus gives the derivative of the function $Af(x)$ (i.e., $A \times f(x)$), where A is any fixed number (i.e., a constant). The derivative is simply A times the derivative of $f(x)$, or the function $Af'(x)$. For example, the derivative of the function $41x^2$ is $41 \times 2x$, which simplifies to $82x$.

Another rule is that the derivative of a sum function of the form $f(x) + g(x)$ is simply the sum of the derivatives of the individual functions, namely $f'(x) + g'(x)$. So, for example, the derivative of the function $x^3 + x^2$ is $3x^2 + 2x$. A similar rule applies for difference functions $f(x) - g(x)$.

Using the above two rules, it is possible to differentiate any polynomial function, since polynomials are built up from powers of x using constant multiples and addition. For example, the derivative of the function $5x^6 - 8x^5 + x^2 + 6x$ is $30x^5 - 40x^4 + 2x + 6$.

In this last example, notice what happens when you differentiate the function $6x$. The derivative is 6 times the derivative of the function x. Applying the rule that takes a power x^n to the derivative nx^{n-1}, the derivative of the function x, which is the same as x^1, is $1x^{1-1}$, that is, $1x^0$. But any number raised to the power 0 is 1. So the derivative of the function x is just 1.

What happens when you try to differentiate a fixed number, say, 11? This problem would arise if you tried to differentiate the polynomial $x^3 - 6x^2 - 4x + 11$. Remember that differentiation is a process that applies to formulas, not to numbers; it is a method of determining slopes. So in order to differentiate 11, you have to think of it not as a number, but as a function—the function that gives the value of 11 for any value of x. Thinking of a number in this way may seem strange, but when drawn as a graph, it is perfectly natural. The 'function' 11 is just a horizontal line drawn 11 units above the x-axis, that is, drawn through the point 11 on the y-axis. You don't need calculus to figure out the slope of this function: it is 0. In other words, the derivative of a constant function, such as the function 11, is 0.

It was in order to provide a foundation for rules of the calculus such as the ones just described that Cauchy developed his theory of limits. In both the case of multiplication by a fixed number and the case of the sum or difference of two functions, the rules (or patterns) of differentiation turn out to be very straightforward. In the case of multiplication of one function by another, the pattern is a little more complicated. The formula for the derivative of a function of the form $f(x)g(x)$ is

$$f(x)g'(x) + g(x)f'(x).$$

For instance, the derivative of the function $(x^2 + 3)(2x^3 - x^2)$ is

$$(x^2 + 3)(6x^2 - 2x) + (2x^3 - x^2)(2x + 0).$$

Other functions for which the derivatives have simple patterns are the trigonometric functions: the derivative of $\sin x$ is $\cos x$, the derivative of $\cos x$ is $-\sin x$, and the derivative of $\tan x$ is $1/(\cos x)^2$.

Even simpler is the pattern for the exponential function: the derivative of e^x is just e^x itself, which means that the exponential function has the unique property that its slope at any point is exactly equal to the value at that point.

In the cases of the sine, cosine, and exponential functions (though not for all functions defined by infinite series), it turns out that the derivative can be obtained by differentiating the infinite series term by term, as if it were a finite polynomial. If you do this, you will be able to verify the above differentiation results for yourself.

Differentiation of the natural logarithm function $\ln x$ also produces a simple pattern: the derivative of $\ln x$ is the function $1/x$.

Is there a danger from the radiation?

In 1986, at Chernobyl in the Ukraine, a disaster at a nuclear power plant caused the release of radioactive material into the atmosphere. The authorities claimed that the amount of radioactivity in the surrounding areas would at no stage reach a catastrophically dangerous level. How could they arrive at this conclusion? More generally, under such circumstances, how can you predict what the level of radioactivity will be a day or a week in the future, in order that any necessary evacuations or other precautions can be carried out?

The answer is, you solve a differential equation—an equation that involves one or more derivatives.

Suppose you want to know the amount of radioactivity in the atmosphere at any time t after the accident. Since radioactivity varies over time, it makes sense to write this as a function of time, $M(t)$. Unfortunately, when you start the investigation, you probably do not have a formula with which to calculate the value at any given time. However, physical theory leads to an equation that connects the rate of increase of radioactive material, dM/dt, with the constant rate, k, at which the radioactive material is released into the atmosphere and the constant rate, r, at which the radioactive material decays. The equation is

$$\frac{dM}{dt} = \frac{rk}{r - M}.$$

This is an example of a differential equation, an equation that involves one or more derivatives. Solving such an equation means finding a formula for the unknown function $M(t)$. Depending on the equation, this may or may not be possible. In the case of the radioactive contamination scenario just described, the equation is particularly simple, and a solution can be obtained. The solution is the function

$$M(t) = \frac{k}{r(1 - e^{-rt})}.$$

When you draw a graph of this function—the first of the four shown in Figure 3.6—you see that the curve rises rapidly at first, but then gradually levels off, getting closer to, but never reaching, the limiting value k/r. Thus the highest level the contamination will ever reach is no greater than k/r.

The same kind of differential equation arises in many other circumstances: in physics, for example, in Newton's law of cooling; in psychology, as a result of studies of learning (the so-called Hullian learning curve); in medicine, in which it describes the rate of intravenous infusion of medication; in sociology, in measurement of the spread of information by mass media; and in economics, in the phenomena of depreciation, sales of new products, and growth of a business. The overall pattern in which some quantity is growing toward a maximum value, is called *limited growth*.

In general, a differential equation arises whenever you have a quantity that is subject to change, and theory provides you with a growth pattern in the form of an equation. Strictly speaking, the changing quantity should be one that changes continuously, which means that it can be captured by means of a function of a real-number variable. Change in many real-life situations, however, consists of a large number of indi-

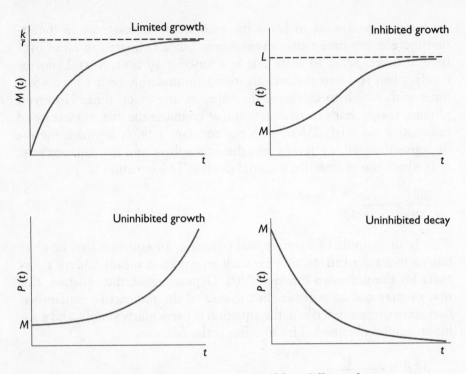

FIGURE **3.6** Solutions to differential equations of four different forms.

vidual, discrete changes that are minuscule compared with the overall scale of the problem, and in such cases there is no harm in simply assuming that the whole changes continuously. This assumption enables the full power of the calculus to be brought to bear in order to solve the differential equation that results. Most applications of differential equations in economics are of this nature: the actual changes brought about in an economy by single individuals and small companies are so small compared with the whole, and there are so many of them, that the whole system behaves as if it were experiencing continuous change.

Other kinds of change give rise to differential equations of other forms. For example, the differential equation

$$\frac{dP}{dt} = rP$$

describes what is called *uninhibited growth*, where $P(t)$ is the size of some population and r is a fixed growth rate. The solution function in this case is

$$P(t) = Me^{rt},$$

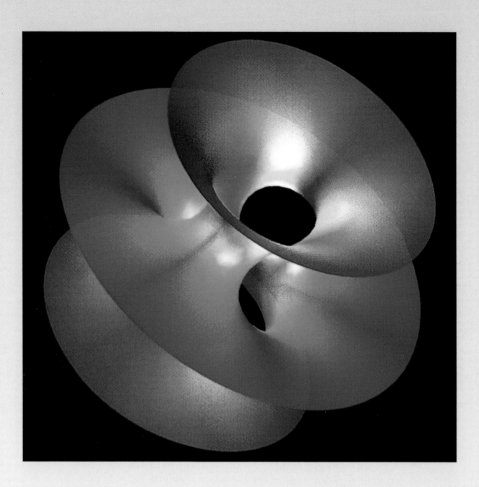

PLATE 1 A minimal surface, discovered by David Hoffman and William Meeks III in 1983.

PLATE 2   Semitic tribe asks permission to enter Egypt. This A.D.-nineteenth-century copy of a nineteenth-century-B.C. mural from the tomb of Beni-Hassan is typical of pre-Renaissance art in its lack of any sense of depth. All the figures appear flat and two-dimensional. (Erich Lessing/Art Resource, New York, N.Y.)

PLATE 3   *The Annunciation*, the Master of the Barberini, 1450. This painting from the Renaissance period exhibits single-point perspective, in which all the perspective lines converge to a single 'point at infinity'.

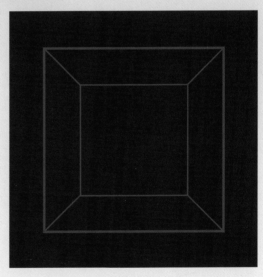

PLATE 4   *Left:* A photo of a physical model of a three-dimensional 'cube-on view' of a hypercube. *Right:* A two-dimensional, face-on view of a regular cube.

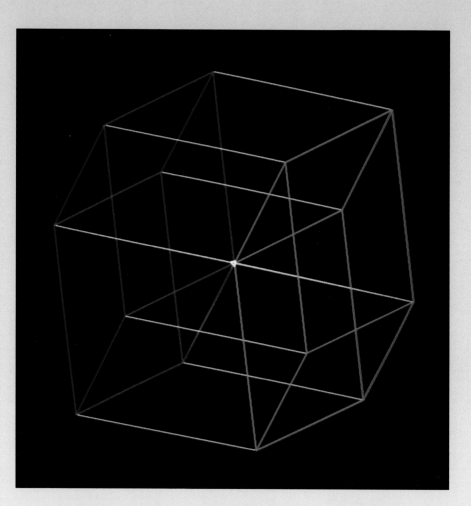

PLATE 5 A still frame from the 1978 movie *The Hypercube: Projections and Slicings,* produced on a computer by Thomas Banchoff and Charles Strauss at Brown University. One of the earliest attempts to use computer graphics to visualize complex mathematical objects, this movie won an international award.

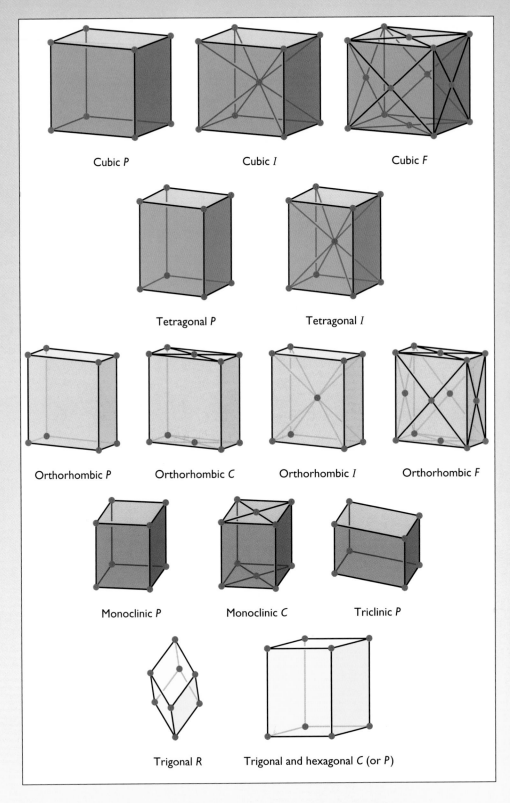

Cubic *P*

Cubic *I*

Cubic *F*

Tetragonal *P*

Tetragonal *I*

Orthorhombic *P*

Orthorhombic *C*

Orthorhombic *I*

Orthorhombic *F*

Monoclinic *P*

Monoclinic *C*

Triclinic *P*

Trigonal *R*

Trigonal and hexagonal *C* (or *P*)

PLATE 6   In 1848, Auguste Bravais proved that there are exactly **fourteen distinct kinds of three-dimensional lattices.**

PLATE 7   This textile design by William Morris exhibits clear translational symmetry.

PLATE 8 Examples of wallpaper patterns of the seventeen distinct possible symmetry groups.

where M is the initial size of the population. The graph of this solution is shown on the top right in Figure 3.6. Over the short term, animal populations, epidemics, and cancers can grow according to this pattern, as can inflation.

Over longer periods, a far more likely scenario than uninhibited growth is that of *inhibited growth*, which is captured by the differential equation

$$\frac{dP}{dt} = rP(L - P),$$

where L is some limiting value for the population. This equation has the solution function

$$P = \frac{ML}{M + (L - M)e^{-Lrt}}.$$

If you draw the graph of such a function, as in the third graph in Figure 3.6, you see that it begins at the initial value M, grows slowly at first, then starts to grow more rapidly until it nears the limiting value L, when the rate of growth steadily slows down.

Finally, the differential equation

$$\frac{dP}{dt} = -rP$$

describes what is called *uninhibited decay*. The solution function is

$$P(t) = Me^{-rt}.$$

Radioactive decay and the depletion of certain natural resources follow this pattern, which is also graphed in Figure 3.6.

More complex forms of differential equations often involve the derivative of the derivative, generally called the *second derivative*. This is particularly true of many differential equations that arise in physics.

The task of finding solutions to differential equations is an entire branch of mathematics in its own right. In many cases, it is not possible to obtain a solution given by a formula; instead, computational methods are used to obtain numerical or graphical solutions.

Because differential equations arise in almost every walk of life, their study is a branch of mathematics that has enormous consequences for humanity. Indeed, from a quantitative point of view, differential equations describe the very essences of life: growth, development, and decay.

Waves that drive the pop music industry

Though they may not know it, when today's pop groups use a music synthesizer, the sounds they produce depend on mathematics developed by a group of eighteenth-century European mathematicians. Today's music synthesizer, which uses computer technology to create complex sounds from the pure notes produced by simple oscillator circuits, is a direct consequence of both the calculus and the development of techniques for manipulating infinite series. Though the technology is very recent, the mathematical theory behind it was worked out in the late eighteenth century by Jean d'Alembert, Daniel Bernoulli, Leonhard Euler, and Joseph Fourier. It is known as *Fourier analysis*, and it deals not with infinite series of numbers, but infinite series of functions.

A striking application of this theory is that, in principle, given enough tuning forks, you could give a performance of Beethoven's Ninth Symphony, complete in every way, including the choral part. (In practice, it would take a very large number of tuning forks indeed to create the complex sounds normally produced by brass, woodwinds, strings, percussion, and human voices. But, in principle, it can be done.)

The crux of the matter is that any sound wave, such as the one shown at the top of Figure 3.7, or indeed, any wave of any kind, can be obtained by adding together an infinite series of *sine waves*—pure waveforms—as shown at the bottom of Figure 3.7. (A tuning fork produces a sound whose waveform is a sine wave.) For example, the three sine waves in Figure 3.8 add together to give the more complex wave shown beneath them. Of course, this is a particularly simple example. In practice, it may take a great many individual sine waves to give a particular waveform; mathematically, it may require an infinite number.

The mathematical result that describes how a wave can be split up into a sum of sine waves is called *Fourier's theorem*. It can be applied to any phenomenon, such as a sound wave, that can be considered as a function of time that keeps on repeating some cycle of values—what is called a *periodic function*. The theorem says that if y is such a periodic function

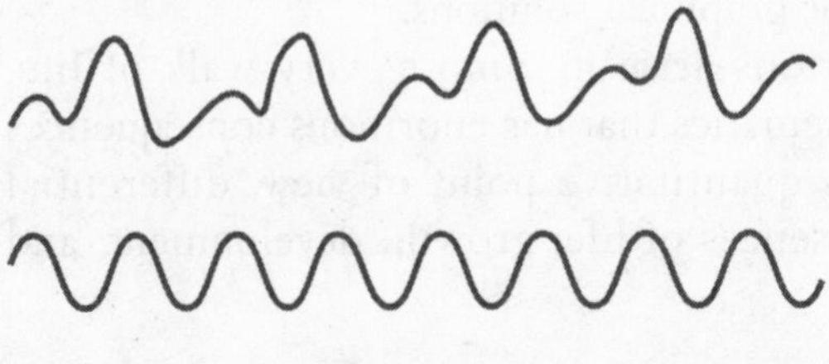

FIGURE **3.7** *Top:* A typical sound wave. *Bottom:* A sine wave.

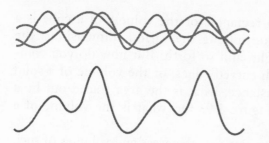

FIGURE **3.8** The three waves at the top sum to form the wave below.

of time, and if the frequency with which y cycles through its period is, say, 100 times a second, then y can be expressed in a form such as

$$y = 4 \sin 200\pi t + 0.1 \sin 400\pi t + 0.3 \sin 600\pi t + \ldots$$

This sum may be finite, or it may continue indefinitely to give an infinite series. In each term, the time t is multiplied by 2π times the frequency. The first term is called the first harmonic, and its frequency is called the fundamental frequency (100 in the example given). The other terms are called higher harmonics, and all have frequencies that are exact multiples of the fundamental frequency. The coefficients (4, 0.1, 0.3, and so on) must all be adjusted to give the particular waveform y. Determination of these coefficients from observed values of the function y, using various techniques from the calculus, constitutes what is known as the Fourier analysis of y.

In essence, Fourier's theorem tells us that the pattern of any sound wave, or indeed, of any kind of wave, no matter how complex, can be built up from the simple, pure wave pattern produced by the sine function. Interestingly, Fourier did not prove this result. He did formulate the theorem, however, and—using some very dubious reasoning that would certainly not be accepted as valid today—gave an argument to show that it was likely to be true. By naming the technique in his honor, the mathematical community acknowledged that the most significant step was to identify the pattern.

Making sure it all adds up

The development of the differential calculus brought with it a surprising bonus, one that was scarcely to be expected: it turns out that the fundamental patterns of differentiation are the same as the patterns that underlie the computation of areas and volumes. More precisely, the computation of areas and volumes is, essentially, the inverse of differentiation—the process of finding slopes. This amazing observation is the basis of an entire second branch of the calculus, the *integral calculus*.

Computing the area of a rectangle or the volume of a cube is a straightforward matter of multiplying the various dimensions: the length, the breadth, the height, and so forth. But how do you compute the area of a figure with curved edges, or the volume of a solid with curved surfaces? For instance, what is the area traced out by a parabola, shown shaded in Figure 3.9? Or, what is the volume of a cone?

The first known attempt to compute the areas and volumes of such geometric figures was made by Eudoxus, a student of Plato at the Academy in Athens. Eudoxus invented a powerful, and exceedingly clever, method of computing areas and volumes, known as the method of exhaustion. Using this method, he was able to show that the volume of any cone is equal to one-third the volume of a cylinder having the same base and equal height, a remarkable pattern that is neither obvious nor easy to prove (see Figure 3.10).

Archimedes used Eudoxus' method to compute the areas and volumes of a number of figures; for example, he found the area traced out by a parabola. This example serves to illustrate how the method of exhaustion works. The idea is to approximate the curve by a series of straight lines, as shown in Figure 3.11. The area traced out by the straight lines consists of two triangles (at the ends) and a number of trapeziums. Since there are simple formulas for the area of a triangle and the area of a trapezium, you can compute the area traced out by the straight lines by simply adding together the areas of these figures. The answer you get will be an approximation to the area beneath the parabola. You can obtain a better approximation by increasing the number of straight lines and repeating the calculation. The method of exhaustion works by gradually increasing the number of straight lines, giving better and better approximations to the area beneath the parabola. You stop when you feel the approximation is good enough.

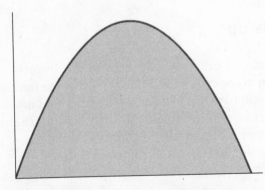

FIGURE **3.9** The area traced out by a parabola.

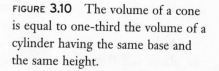

FIGURE **3.10** The volume of a cone is equal to one-third the volume of a cylinder having the same base and the same height.

This process is called the method of exhaustion, not because Eudoxus became exhausted after computing so many approximations, but because the sequence of successively approximated areas would eventually exhaust the entire area beneath the original curve if continued indefinitely.

Interest in geometric figures such as parabolas and ellipses was revived during the early part of the seventeenth century, when Johannes Kepler observed three elegant, and profound, mathematical patterns in the sky—his now-famous laws of planetary motion: (1) a planet orbits the sun in an ellipse having the sun at one of its two foci; (2) a planet

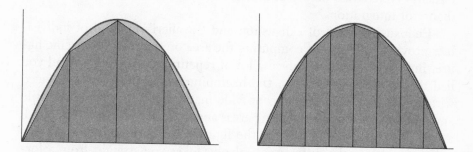

FIGURE **3.11** The area beneath the parabola is approximated by the sum of the triangles and trapeziums. The greater the number of subdivisions, the more accurate the approximation.

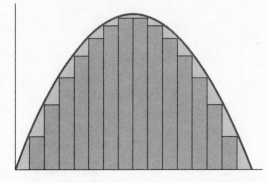

FIGURE **3.12**   The method of indivisibles.

sweeps out equal areas in equal times; and (3) the cube of a planet's distance from the sun is equal to the square of its orbital period.

Mathematicians of the time, among them Galileo, Kepler himself, and above all, Bonaventura Cavalieri of Italy, computed areas and volumes by means of the *method of indivisibles*. In this approach, a geometric figure is regarded as being made up of an infinite number of 'atoms' of area or volume, which are added together to give the required area or volume. The general idea is illustrated in Figure 3.12. Each of the shaded areas is a rectangle, whose area can be calculated precisely. When there are only a finite number of such rectangles, as shown, adding together all these areas gives an approximation to the area under the parabola. If there were infinitely many rectangles, all infinitesimally wide, their addition would produce the true area—if only it were possible to carry out this infinite computation. In his book *Geometria Indivisibilus Continuorum*, published in 1635, Cavalieri showed how to handle indivisibles in a reasonably reliable way in order to come up with the right answer. When placed on a rigorous footing with the aid of the Cauchy–Weierstrass theory of limits, this approach became the modern theory of integration.

Eudoxus' method of exhaustion and Cavalieri's method of indivisibles provided a means of computing the area or volume of a specific figure. But each method involved a lot of repetitive computation, and you had to start over again from the beginning each time you were faced with a new figure. In order to provide mathematicians with a versatile and efficient means of computing areas and volumes, more was required: a way to go from a formula for the figure to a formula for the area or volume, much as the differential calculus takes you straight from a formula for a curve to a formula for its slope.

This is where a truly amazing thing happens. Not only is there such a general method, it turns out to be a direct consequence of the method of differentiation. As with the differential calculus, the key step is to consider not the problem of computing a particular area or volume, but the more general task of finding an area or volume *function*.

Take the case of computing areas. The curve shown in Figure 3.13 traces out an area. More precisely, it determines an area function: for any point x, there will be a corresponding area, the area shown shaded in the picture. (This is a specially chosen, simple case. The general situation is a bit more complicated, but the essential idea is the same.) Let $A(x)$ denote this area, and let $f(x)$ be the formula that determines the original curve. In any particular example, you will know the formula for $f(x)$, but you will not have a formula for $A(x)$.

Even if you don't know its formula, $A(x)$ is still a function, and so might have a derivative, in which case you can ask what its derivative is. And, lo and behold, the answer turns out to be none other than the function $f(x)$, the very formula for the curve that traces out the area in the first place.

This is true for any function $f(x)$ given by a reasonable formula, and can be proved *without knowing a formula for $A(x)$!* The proof depends only on the general patterns involved in computing areas and derivatives.

Briefly, the idea is to look at the way the area $A(x)$ changes when you increase x by a small amount h. Referring to Figure 3.14, notice that the new area $A(x + h)$ can be split up into two parts: $A(x)$ plus a small additional area that is very nearly rectangular. The width of this additional rectangle is h; its height is given by reading off from the graph: it is $f(x)$. So the area of the extra piece is $h \times f(x)$ (width times height). The entire area is given by the approximation

$$A(x + h) \approx A(x) + h \times f(x).$$

The symbol $\approx$ denotes approximate equality.

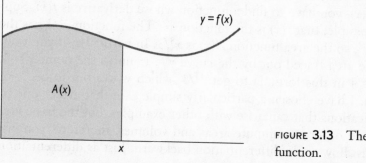

$y = f(x)$

$A(x)$

x

FIGURE **3.13** The area function.

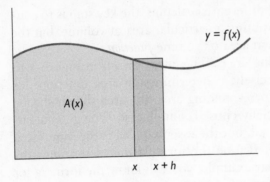

FIGURE **3.14** The proof of
the fundamental theorem of
calculus.

The above equation can be rearranged to look like this:

$$\frac{A(x + h) - A(x)}{h} \approx f(x).$$

Remember, this identity is only approximate, since the additional area you have to add to $A(x)$ to give the area $A(x + h)$ is not exactly rectangular. But the smaller h is, the better this approximation will be.

The expression on the left looks familiar, doesn't it? It is precisely the expression that gives the derivative $A'(x)$ when you evaluate the limit as h approaches 0. So, as h becomes smaller and smaller, three things happen: the equation becomes more and more accurate, the expression on the left approaches the limit $A'(x)$, and the right-hand side remains constant at the value $f(x)$, whatever that value is. You finish up with, not an approximation, but the genuine identity

$$A'(x) = f(x).$$

This remarkable result, connecting the tasks of computing slopes and computing areas (it also works for volumes), is known as the *fundamental theorem of calculus.*

The fundamental theorem of calculus provides a method of finding a formula for $A(x)$. In order to find the area function $A(x)$ for a given curve $y = f(x)$, you have to find a function whose derivative is $f(x)$. Suppose, for example, that $f(x)$ is the function x^2. The function $x^3/3$ has the derivative x^2, so the area function $A(x)$ is $x^3/3$. In particular, if you want to know the area traced out by the curve $y = x^2$ up to the point $x = 4$, you set $x = 4$ in this formula to get $4^3/3$, which works out to be $\frac{64}{3}$, or $21\frac{1}{3}$. (Again, I have chosen a particularly simple case that avoids one or two complications that can arise with other examples, but the basic idea is correct.) In order to compute areas and volumes, therefore, you just have to learn how to do differentiation backward. Just as differentiation itself is so routine that it can be programmed into a computer, there are computer programs that can perform integration.

The fundamental theorem of calculus stands as a shining example of the huge gain that can result from a search for deeper, more general, and more abstract patterns. In the cases of finding slopes and of computing areas and volumes, the ultimate interest may well be in finding a particular number, and yet the key in both cases is to look at the more general, and far more abstract, patterns whereby the slope and the area or volume change with varying x values.

The real numbers

The question of whether or not time and space are continuous or have a discrete, atomic nature has had important implications since the beginning of science. Practically all of the scientific and mathematical developments that took place from Greek times up to the end of the nineteenth century were built on the assumption that time and space are continuous and not discrete. Regarding both time and space as a continuum was assumed to avoid the paradoxes of Zeno. This view had been prevalent since the time of Plato, who proposed that the continuum arose as a result of the flowing of the *apeiron*, a highly abstract entity that does not really have a counterpart in any present-day theories.

By the time of Newton and Leibniz, the continuum that arose from the physical world of time and space was equated with a continuum of what were called *real numbers*. Numerical measurements of time and of physical quantities, such as length, temperature, weight, velocity, and so on, were assumed to be points on this continuum. The differential calculus applied to functions involving variables that ranged over the real-number continuum.

In the 1870s, when Cauchy, Weierstrass, Richard Dedekind, and others tried to develop a theory of limits adequate to support the techniques of the calculus, they had to carry out a deep investigation of the nature of the real-number continuum. Their starting point was to regard that continuum as a set of points—the real numbers—arranged in a line that stretches out to infinity in both directions.

The real numbers are an extension of the rational numbers, and many of the axioms for the real numbers are also axioms for the rationals. In particular, this is the case for the arithmetical axioms that specify properties of addition, subtraction, multiplication, and division. The arithmetical axioms ensure that the real numbers are a field (see page 73). There are also axioms that describe the ordering of the real numbers, and again, these axioms are also axioms for the rational numbers.

The key axiom that distinguishes the real numbers from the rational numbers is the one that allows the development of an adequate theory of limits. Though the rational numbers have all of the necessary arithmetical and order properties needed for the calculus, they are not at all suited to a theory of limits. As formulated by Cauchy, this additional axiom reads as follows:

> Suppose that $a_1, a_2, a_3, \ldots$ is an infinite sequence of real numbers that get closer and closer together (in the sense that, the further along the sequence you go, the differences between the numbers become arbitrarily close to 0). Then there must be a real number, call it L, such that the numbers in the sequence get closer and closer to L (in the sense that, the further along the sequence you go, the differences between the numbers a_n and the number L become arbitrarily close to 0).

The number L is known as the *limit* of the sequence $a_1, a_2, a_3, \ldots$.

Notice that the rational numbers do not have this property. The sequence 1, 1.4, 1.41, 1.414, ... of successive rational approximations to $\sqrt{2}$ is such that the terms get closer and closer together, but there is no single *rational* number L such that the terms in the sequence get arbitrarily close to L. (The only possibility for the number L is $\sqrt{2}$, which we know is not rational.)

Cauchy's axiom is known as the *completeness axiom*. Cauchy gave a formal construction of the real numbers by starting with the rational numbers and adding (using a sophisticated method) new points to the rational number line. The new points were defined to be the limits of all sequences of rationals that have the property of getting closer and closer together. An alternative construction of the real numbers from the rationals was given by Dedekind.

The construction of the real numbers and the development of rigorous theories of limits, derivatives, and integrals, begun by Cauchy, Dedekind, Weierstrass, and others, was the beginning of the subject nowadays referred to as *real analysis*. These days, a fairly extensive study of real analysis is regarded as an essential component of any college education in mathematics.

Complex numbers

That what started out as an investigation of motion and change should lead to theories of limits and of the real continuum is perhaps not surprising, given the two-thousand-year-old paradoxes of Zeno. Far more surprising is that the invention of the calculus also led to the acceptance into the mainstream of mathematics of a system of numbers that includes such a counterintuitive entity as the square root of -1. And yet this is exactly what happened, with Cauchy as one of the prime movers in the development.

The story begins about a hundred years before Newton and Leibniz carried out their work. European mathematicians of the sixteenth century, in particular the Italians Girolamo Cardano and Rafaello Bombelli, began to realize that, in trying to solve algebraic problems, it was sometimes useful to assume the existence of negative numbers, and moreover, to assume that negative numbers have square roots. Both assumptions were widely regarded as extremely dubious, at worst utter nonsense and at best having solely a utilitarian purpose.

Since the time of the ancient Greeks, mathematicians had known how to manipulate expressions involving minus signs, using rules such as $-(-a) = a$ and $1/-a = -1/a$. However, they felt this was permissible only if the final answer was positive. Their distrust of negative numbers was largely a legacy of the Greek notion of numbers representing lengths and areas, which are always positive. Not until the eighteenth century were negative numbers accepted as bona fide numbers.

It took even longer to accept that the square root of a negative number can be a genuine number. The reluctance to accept these numbers is reflected in the use of the term *imaginary number* for such an entity. As with negative numbers, mathematicians allowed themselves to manipulate imaginary numbers during the course of a calculation. Indeed, arithmetic expressions involving imaginary numbers can be manipulated using the ordinary rules of algebra. The question was, do such numbers exist?

This question can be reduced to the existence or otherwise of a single imaginary number, the square root of -1. Here's why. Suppose there is such a number as $\sqrt{-1}$. Following Euler, denote it by the letter i, for imaginary. Then the square root of any negative number $-a$ is simply $i\sqrt{a}$, the product of this special number i and the square root of the positive number a.

Ignoring the question of whether the number i really exists, mathematicians introduced hybrid numbers of the form $a + bi$, where a and b are real numbers. These hybrids are called *complex numbers*. Using the ordinary rules of algebra, together with the fact that $i^2 = -1$, it is possible to add, subtract, multiply, and divide complex numbers; for instance:

$$(2 + 5i) + (3 - 6i) = (2 + 3) + (5 - 6)i$$
$$= 5 - 1i$$
$$= 5 - i$$

and (using dots to denote multiplication):

$$(1 + 2i)(3 + 5i) = 1 \cdot 3 + 2 \cdot 5 \cdot i^2 + 2 \cdot 3 \cdot i + 1 \cdot 5 \cdot i$$
$$= 3 - 10 + 6i + 5i$$
$$= -7 + 11i.$$

(Division is a little more complicated.)

In present-day terminology, the complex numbers would be said to constitute a field, just as do the rationals and the reals. Unlike the rationals and the reals, however, the complex numbers are not ordered—there is no natural notion of 'greater than' for complex numbers. Instead of being points on a line, as are the rationals and the reals, the complex numbers are points in the complex plane: the complex number $a + bi$ is the point with coordinates a and b, as shown in Figure 3.15.

In the complex plane, the horizontal axis is referred to as the *real axis* and the vertical axis is referred to as the *imaginary axis*, since all real numbers lie on the horizontal axis and all imaginary numbers lie on the vertical axis. All other points in the complex plane represent complex numbers, which are the sum of a real number and an imaginary number.

Since the complex numbers are not points on a line, you cannot say which one of two complex numbers is the larger; for the complex numbers there is no such notion. There is, however, a notion of size of sorts. The absolute value of a complex number $a + bi$ is the distance from the origin to the number, measured in the complex plane; it is usually denoted by $|a + bi|$. By the Pythagorean theorem,

$$|a + bi| = \sqrt{a^2 + b^2}.$$

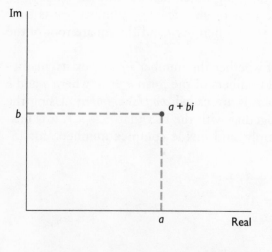

FIGURE **3.15** The complex plane.

The absolute values of two complex numbers may be compared, but it is possible for different complex numbers to have the same absolute value; for example, both $3 + 4i$ and $4 + 3i$ have the absolute value 5.

Complex numbers might seem, on the face of it, to be just a curiosity devised by mathematicians for their own amusement. The reality is far from that. You get your first hint that something deep is going on when you discover that complex numbers have deep and significant implications for the solution of polynomial equations.

Where all equations can be solved

The natural numbers are the most basic number system of all. Though useful for counting, they are not suited to solving equations. Using the natural numbers, it is not possible to solve even such a simple equation as

$$x + 5 = 0.$$

In order to solve equations of this kind, you need to go to the integers.

But the integers are also impoverished, in that they do not allow you to solve a simple linear equation such as

$$2x + 3 = 0.$$

In order to solve equations of this kind, you need to go to the rationals.

The rationals are adequate to solve all linear equations, but do not allow you to solve all quadratic equations; for example, the equation

$$x^2 - 2 = 0$$

cannot be solved using the rationals.

The real numbers are sufficiently rich to solve this quadratic equation. But the reals do not allow you to solve all quadratic equations; for example, you cannot solve the equation

$$x^2 + 1 = 0$$

using the reals. To solve this quadratic equation, you need to go to the complex numbers.

At this point, primed as you are to look for patterns in mathematics, you might be tempted to suppose that this process goes on forever: each time you move to a richer number system, you find yet another kind of equation that you cannot solve. But this is not the case. When you reach the complex numbers, the process comes to a halt. Any polynomial equation

$$a_n x^n + a_{n-1} x^{n-1} + \ldots + a_1 x + a_0 = 0$$

where the coefficients $a_0, \ldots, a_n$ are complex numbers can be solved in the complex numbers.

This important result is known as the *fundamental theorem of algebra*. It was suspected, but not proved, in the early seventeenth century. Incorrect proofs were supplied by d'Alembert in 1746 and by Euler in 1749. The first correct proof was given by Gauss in his 1799 doctoral thesis.

Euler's amazing formula

Complex numbers turn out to have connections to many other parts of mathematics. A particularly striking example comes from the work of Euler. In 1748, he discovered the amazing identity

$$e^{ix} = \cos x + i \sin x,$$

which is true for any real number x.

Such a close connection between trigonometric functions, the mathematical constant e, and the square root of -1 is already quite startling. Surely such an identity cannot be a mere accident; rather, we must be catching a glimpse of a rich, complicated, and highly abstract mathematical pattern that for the most part lies hidden from our view.

In fact, Euler's formula has other surprises in store. If you substitute the value π for x in Euler's formula, then, since $\cos \pi = -1$ and $\sin \pi = 0$, you get the identity

$$e^{i\pi} = -1.$$

Rewriting this as

$$e^{i\pi} + 1 = 0,$$

you obtain a simple equation that connects the five most common constants of mathematics: e, π, i, 0, and 1.

Not the least surprising aspect of this last equation is that the result of raising an irrational number to a power that is an irrational imaginary number can turn out to be a natural number. Indeed, raising an imaginary number to an imaginary power can also give a real-number answer. Setting $x = \pi/2$ in the first equation, and noting that $\cos \pi/2 = 0$ and $\sin \pi/2 = 1$, you get

$$\frac{e^{i\pi}}{2} = i,$$

and, if you raise both sides of this identity to the power i, you obtain (since $i^2 = -1$)

$$e^{-\pi/2} = i^i.$$

Thus, using a calculator to compute the value of $e^{-\pi/2}$, you find that

$$i^i = 0.207\ 879\ 576 \ldots$$

(I should mention that this is just one of a whole infinity of possible values for the quantity i^i. With complex numbers, exponentiation does not always give a single answer.)

With the gradual increase in the use of complex numbers spurred on by the obvious power of the fundamental theorem of algebra and the elegance of Euler's formula, complex numbers began their path toward acceptance as bona fide numbers. That finally occurred in the middle of the nineteenth century, when Cauchy and others started to extend the methods of the differential and integral calculus to include the complex numbers. Their theory of differentiation and integration of complex functions turned out to be so elegant—far more than in the real-number case—that on aesthetic grounds alone, it was, finally, impossible to resist any longer the admission of the complex numbers as fully paid-up members of the mathematical club. Provided it is correct, mathematicians never turn their backs on beautiful mathematics, even if it flies in the face of all their past experience.

In addition to its mathematical beauty, however, complex calculus—or complex analysis, as it is referred to these days—turned out to have significant applications to, of all things, the theory of the natural numbers. The finding that there was a deep and profound connection between complex analysis and the natural numbers was yet another testimony to the power of mathematical abstraction. The techniques of complex calculus enabled number theorists to identify and describe number patterns that, in all likelihood, would otherwise have remained hidden forever.

Uncovering the hidden patterns of numbers

The first person to use the methods of complex calculus to study properties of the natural numbers—a technique known nowadays as *analytic number theory*—was the German mathematician Bernhard Riemann. In a paper published in 1859, titled *On the number of primes less than a given magnitude*, Riemann used complex calculus to investigate a number-theoretic pattern that had first been observed by Gauss: for large natural numbers N, the number of primes less than N, denoted by $\pi(N)$, is approximately equal to the ratio $N/\ln N$ (see page 27 for a table of values of the function $\pi(N)$). Since both $\pi(N)$ and $N/\ln N$ grow increasingly large as N increases, you have to formulate this observation with some care. The precise formulation is that the limit of the ratio $\pi(N)/[N/\ln N]$ as N approaches infinity is exactly equal to 1. This observation was known as the *prime number conjecture*.

Prior to Riemann, the closest anyone had come to a proof of the prime number conjecture was a result obtained by Pafnuti Chebyshev in 1852, which said that, for sufficiently large values of N, $\pi(N)/[N/\ln N]$ lies between 0.992 and 1.105. In order to obtain this result, Chebyshev made use of a function introduced by Euler back in 1740, called the zeta function after the Greek letter Euler had used to denote it.

Euler defined the zeta function by the infinite series

$$\zeta(x) = \frac{1}{1^x} + \frac{1}{2^x} + \frac{1}{3^x} + \frac{1}{4^x} + \dots$$

The number x here can be any real number greater than 1. If x is less than or equal to 1, this infinite series does not have a finite sum, so $\zeta(x)$ is not defined for such x. If you set $x = 1$, the zeta function gives the harmonic series, considered earlier in the chapter. For any value of x greater than 1, the series yields a finite value.

Euler showed that the zeta function is related to the prime numbers by proving that, for all real numbers x greater than 1, the value of $\zeta(x)$ is equal to the infinite product

$$\frac{1}{1 - \left(\frac{1}{2}\right)^x} \times \frac{1}{1 - \left(\frac{1}{3}\right)^x} \times \frac{1}{1 - \left(\frac{1}{5}\right)^x} \times \dots$$

where the product is over all numbers of the form

$$\frac{1}{1 - \left(\frac{1}{p}\right)^x}$$

where p is a prime number.

This connection between the infinite series of the zeta function and the collection of all prime numbers is already quite striking—after all, the primes seem to crop up among the natural numbers in a fairly haphazard way, with little discernible pattern, whereas the infinite series of the zeta function has a very clear pattern, progressing steadily up through all the natural numbers one at a time.

The major step taken by Riemann was to show how to extend the definition of the zeta function to give a function $\zeta(z)$ defined for all complex numbers z. (It is customary to denote a complex variable by the letter z, just as x is commonly used to denote a real variable.) To achieve his result, Riemann used a complicated process known as analytic continuation. This process works by extending a certain abstract pattern possessed by Euler's function, though it is a pattern whose abstraction is well beyond the scope of a book such as this.

Why did Riemann go to all this effort? Because he realized that he could prove the prime number conjecture if he was able to determine the complex solutions to the equation

$$\zeta(z) = 0.$$

The complex solutions to this equation are generally referred to as the *complex zeros* of the zeta function. (Notice that this is a technical use of the word *zero*. In this sense, the 'zeros' are the values of z for which the function $\zeta(z)$ is equal to 0.)

The real zeros of the zeta function are easy to find: they are -2, -4, -6, and so on, the negative even integers. (Remember, Euler's definition of the zeta function in terms of an infinite series works only for real numbers x greater than 1. I am now talking about Riemann's extension of this function.)

In addition to these real zeros, the zeta function also has infinitely many complex zeros. They are all of the form $x + iy$, where x lies between 0 and 1; that is, in the complex plane, they all lie between the y-axis and the vertical line $x = 1$. But can we be more precise than this? In his paper, Riemann proposed the hypothesis that all the zeros other than the negative even integers are of the form $\frac{1}{2} + iy$; that is, they lie on the line $x = \frac{1}{2}$ in the complex plane. The prime number conjecture follows from this hypothesis.

Riemann must have based this hypothesis on his understanding of the zeta function and the pattern of its zeros. He certainly had little by way of numerical evidence—that came only much later, with the advent of the computer age. Computations carried out over the last thirty years have shown that the first one and a half billion zeros all lie on the appropriate line. But despite this impressive-looking numerical evidence, the Riemann hypothesis remains unproved to this day. Most mathematicians would agree that it is *the* most significant unsolved problem of mathematics.

The prime number conjecture was finally proved by Jacques Hadamard, and independently by Charles de la Vallée Poussin, in 1896.

Their proofs used the zeta function, but did not require Riemann's hypothesis.

When the prime number conjecture was proved—becoming the *prime number theorem*—mathematicians had come full circle. Mathematics began with the natural numbers, the building blocks of counting. With the invention of the calculus by Newton and Leibniz, mathematicians came to grips with the infinite, and were able to study continuous motion. The introduction of the complex numbers and the proof of the fundamental theorem of algebra provided the facility to solve all polynomial equations. Then Cauchy and Riemann showed how to extend the calculus to work for complex functions. Finally, Riemann and others used the resulting theory—a theory of considerable abstraction and complexity—to establish new results about the natural numbers.

CHAPTER 4

Mathematics Gets into Shape

Everyone is a geometer

What do you see in the diagram in Figure 4.1? At first glance, like everyone else, you probably see a triangle. But look more closely, and you will see that there is no triangle on the page, merely a collection of three black disks, with a bite missing out of each. The triangle that you saw is an optical illusion, produced without any conscious effort on your part. Your mind and visual system filled in details such as lines and surfaces in order to obtain geometric cohesion. The human visual–cognitive system constantly looks for geometric patterns. In this sense, we are all geometers.

Having acknowledged that we 'see' geometric shapes, what is it that enables you to recognize a triangle as a triangle, be it on the page, in the landscape around you, or in your mind, as in the example shown? Obviously not the size. Nor the color. Nor the thickness of the lines. Rather, it is the shape. Whenever you see three straight lines joined at their ends to form a closed figure, you recognize that figure as a triangle. You do this because you possess the abstract concept of a triangle. Just as the abstract concept of the number 3 transcends any particular collection of three objects, so too the abstract concept of a triangle transcends any particular triangle. In this sense, too, we are all geometers.

Not only do we see geometric patterns in the world around us, we seem to have a built-in preference for some of them. One well-known example of such a pattern is captured by the *golden ratio*, a number mentioned at the beginning of Book VI of Euclid's *Elements*.

FIGURE **4.1** An optical illusion of a triangle.

According to the Greeks, the golden ratio is the proportion for the sides of a rectangle that the human eye finds the most pleasing. The rectangular face of the front of the Parthenon has sides whose ratio is in this proportion, and it may be observed elsewhere in Greek architecture.

The value of the golden ratio is $(1 + \sqrt{5})/2$, an irrational number approximately equal to 1.618. It is the number you get when you divide a line into two pieces so that the ratio of the whole line to the longer piece equals the ratio of the longer piece to the shorter. Expressing this algebraically, if the ratio concerned is x:1, as illustrated in Figure 4.2, then x is a solution to the equation

$$\frac{x + 1}{x} = \frac{x}{1},$$

that is,

$$x^2 = x + 1.$$

The positive root of this equation is $x = (1 + \sqrt{5})/2$.

The golden ratio crops up in various parts of mathematics. One well-known example is in connection with the Fibonacci sequence. This is the sequence of numbers you get when you start with 1 and form the next number in the sequence by adding together the two previous ones (except at step 2, where you have only one previous number). Thus, the sequence begins

1, 1, 2, 3, 5, 8, 13, 21, . . .

This sequence captures a pattern that can be observed in many situations involving growth, from the growth of plants to the growth of a computer database. If $F(n)$ denotes the nth number of this sequence,

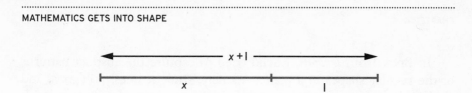

FIGURE **4.2** Dividing a line segment to give the golden ratio.

then, as *n* becomes larger, the ratio $F(n + 1)/F(n)$ of successive terms of the Fibonacci sequence gets closer and closer to the golden ratio.

The golden ratio has a particularly intriguing representation as a fraction that continues forever, namely,

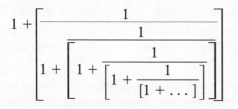

Numerous examples of the golden ratio appear in nature: the relative sizes of the chambers in a nautilus shell, the arrangement of seeds in a sunflower, the pattern you can see on the surface of a pine cone, to name just a few. The fact that we see the golden ratio as somehow pleasing is yet another indication of how our minds are attuned to certain geometric patterns.

Measuring the earth

The word *geometry* comes from the Greek: *geo-metry* means 'earth measurement'. The mathematical ancestors of today's geometers were the land surveyors of ancient Egypt, who had the task of reestablishing boundaries washed away by the periodic flooding of the Nile; the Egyptian and Babylonian architects who designed and built temples, tombs, and the obviously geometric pyramids; and the early seaborne navigators who traded along the Mediterranean coast. Just as these same early civilizations made practical use of numbers without having an obvious concept of number, let alone a theory of such entities, their largely utilitarian use of various properties of lines, angles, triangles, circles, and the like was unaccompanied by any detailed mathematical study.

As mentioned in Chapter 1, it was Thales who, in the sixth century B.C., started the Greek development of geometry as a mathematical discipline—indeed, the first mathematical discipline. Euclid's *Elements*, written around 350 B.C., was largely a book about geometry.

In Book I of *Elements*, Euclid tried to capture the abstract patterns of the regular shapes in a plane—namely, straight lines, polygons, and circles—by means of a system of definitions and postulates (axioms) that was to become known as Euclidean geometry. Among the twenty-three initial definitions he set down are:

Definition 1. A point is that which has no part.

Definition 2. A line is a breadthless length.

Definition 4. A straight line is a line that lies evenly with the points on itself.

Definition 10. When a straight line standing on another straight line makes the adjacent angles equal to one another, each of the equal angles is right and the straight line standing on the other is called a perpendicular to that on which it stands.

Definition 23. Parallel straight lines are straight lines that, being in the same plane and being produced indefinitely in both directions, do not meet one another in either direction.

To the mathematician of today, the first three of the above definitions are unacceptable; they simply replace three undefined notions with other undefined notions, and nothing is gained. In fact, the modern geometer takes the notions of 'point' and 'straight line' as given, and does not attempt to define them. But Euclid's later definitions still make sense.

Notice that the definition of a right angle is entirely nonquantitative; no mention is made of 90° or $\pi/2$. For the Greeks, geometry was nonnumeric, being founded entirely on observation of patterns of shape; in particular, they regarded lengths and angles as geometric notions, not numeric ones.

Having defined, or at least attempted to define, the basic notions, Euclid's next step was to formulate five basic postulates, from which all geometric facts were supposed to follow by means of pure logical reasoning.

Postulate 1. [It is possible] to draw a straight line from any point to any point.

Postulate 2. [It is possible] to produce a finite straight line continuously in a straight line.

Postulate 3. [It is possible] to describe a circle with any center and [radius].

Postulate 4. All right angles are equal to one another.

Postulate 5. If a straight line falling on two straight lines makes the interior angles on the same side less than two right angles, the two straight lines, if produced indefinitely, meet on that side on which are the angles less than the two right angles.

In writing down these postulates, and then deducing other geometric facts from them, Euclid was not trying to establish some kind of logical game with arbitrary rules, like chess. For Euclid, and for generations of mathematicians after him, geometry was the study of regular shapes that can be observed in the world. The five postulates were supposed to be self-evident truths about the world, and in formulating them, Euclid was trying to capture some fundamental patterns of nature.

We live our lives in large part by being able to recognize, and sometimes ascribe, shape. The mathematical study of shape has given rise to several branches of mathematics. Geometry, the most obvious among them, forms the main topic of this chapter. Symmetry and topology, discussed in the following chapters, study different, and in some ways more abstract, patterns of shape.

Things that Euclid missed

Euclid's postulates were intended to provide a basis for the development of plane geometry. In particular, they were intended to capture all that is needed to prove the forty-eight propositions in Book I of *Elements*, culminating in the Pythagorean theorem and its converse. Given such a goal, the five postulates are astonishingly few in number and, in all but one case, extremely simple in content. Unfortunately, they don't express all the assumptions that Euclid used in his proofs. Along with many mathematicians after him, Euclid tacitly assumed a number of facts that he did not formulate as postulates; for example:

- A straight line passing through the center of a circle must intersect the circle.
- A straight line that intersects one side of a triangle, but does not pass through any vertex of the triangle, must intersect one of the other sides.
- Given any three distinct points on the same line, one of them is between the other two.

Since Euclid's aim in developing geometry in an axiomatic fashion was to avoid any reliance on a diagram during the course of a proof, it is perhaps surprising that he overlooked such basic assumptions as those just listed. On the other hand, his was the first serious attempt at axiomatization, and, when compared with the state of, say, physics or medicine in 350 B.C., Euclid's mathematics looks, and was, centuries ahead of its time.

Over two thousand years after Euclid, in the early part of the twentieth century, David Hilbert finally wrote down a list of twenty postulates that is adequate for the development of Euclidean geometry in that all of the theorems in *Elements* can be proved from these postulates, using only pure logic.

The one postulate in Euclid's list that does not have a simple form is the fifth one. Compared with the other four postulates, it is complicated indeed. Stripping away the tangled verbiage, what it says is that if two straight lines in a plane are inclined toward one another, they will eventually meet. Another way to express this is that it is possible to draw exactly one straight line through a given point that is parallel to a given straight line. In fact, the fifth postulate looks more like a theorem than an axiom, and it seems that Euclid himself was reluctant to assume it, since he avoided any use of it in *Elements* until Proposition I.29. Indeed, many subsequent generations of mathematicians attempted to deduce the fifth postulate from the other four, or to formulate a more basic assumption from which the fifth postulate could be deduced.

It was not that anyone doubted the veracity of the postulate. On the contrary, it seems perfectly obvious. It was its logical form that caused the problem: axioms were not supposed to be so specific or complicated. (It is not clear that such a view would prevail today; since the nineteenth century, mathematicians have learned to live with many more complicated axioms that likewise are supposed to capture 'obvious truths'.)

Still, obvious or not, no one was able to deduce the fifth postulate from the others, a failure that was taken to indicate a lack of understanding of the geometry of the world we live in. Nowadays, we recognize that there was indeed a failure of comprehension, but not of Euclidean geometry itself. Rather, the problem lay in the assumption that the geometry Euclid tried to axiomatize *was* the geometry of the world we live in—an assumption that the great philosopher Immanuel Kant, among others, took to be fundamental.

But that story will have to wait until a later section; in the meantime, we shall take a look at some of the results Euclid obtained from his five postulates.

Euclid in his *Elements*

Of the thirteen books that make up *Elements*, the first six are devoted to plane geometry in one form or another.

A number of the propositions in Book I concern ruler-and-compass constructions. The task here is to determine what geometric figures can be constructed using just two tools: an unmarked straightedge, used only to draw straight lines, and a compass, used to draw arcs of circles, but for no other purpose—in particular, the separation of the compass points is assumed to be lost when the instrument is taken from the page. Euclid's very first proposition describes just such a construction:

Proposition I.1. On a given straight line, to construct an equilateral triangle.

The method, illustrated in Figure 4.3, seems simple enough. If the given line is *AB*, place the point of the compass at *A* and draw a quarter-circle above the line with radius *AB*, then place the point of the compass at *B* and draw a second quarter-circle with the same radius. Let *C* be the point where the two quarter-circles intersect. Then *ABC* is the desired triangle.

Even here, in his very first proposition, Euclid makes use of a tacit assumption that his axioms do not support: How do you know that the two quarter-circles intersect? Admittedly, the diagram suggests that they do, but diagrams are not always reliable; perhaps there is a 'hole' at *C*, much like the 'hole' in the rational number line where $\sqrt{2}$ 'ought to be'. In any case, the whole point of writing down axioms in the first place was to avoid reliance on diagrams.

Other ruler-and-compass constructions include the bisection of an angle (Proposition I.9), the bisection of a line segment (Proposition I.10), and the construction of a perpendicular to a line at a point on that line (Proposition I.11).

It should be emphasized that, for all the attention given to them in *Elements*, Greek geometry was by no means restricted to ruler-and-compass constructions. Indeed, Greek mathematicians made use of whatever tools the problem seemed to demand. On the other hand, they did seem to regard ruler-and-compass constructions as a particularly elegant form of

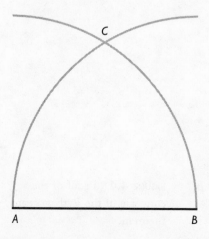

FIGURE **4.3** The construction of an equilateral triangle on a line.

intellectual challenge: to the Greeks, a figure that could be constructed using only these two most primitive of tools was somehow more fundamental and pure, and a solution to a problem using just these tools was regarded as having particular aesthetic appeal. Euclid's postulates are clearly designed to try to capture what can be achieved using ruler and compass.

In addition to the results on constructions, Book I gives a number of criteria for establishing that two triangles are congruent (i.e., equal in all respects). For example, Proposition I.8 states that when the three sides of one triangle are respectively equal to the three sides of another, then the two triangles are congruent.

The final two propositions in Book I are the Pythagorean theorem (Proposition I.47) and its converse. Euclid's proof of the latter is so elegant that I shall present it here. Figure 4.4 provides the diagram.

Proposition I.48. If in a triangle the square on one of the sides be equal to the squares on the remaining two sides of the triangle, the angle contained by the remaining two sides of the triangle is right.

PROOF: Start with triangle ABC, in which it is assumed that $BC^2 = AB^2 + AC^2$. It is to be proved that angle BAC [often written as $\angle BAC$] is right.

To do this, first draw AE perpendicular to AC at A. This step is possible by Proposition I.11. Then construct the segment AD so that $AD = AB$. This is allowed by Proposition I.3.

The aim now is to show that triangles BAC and DAC are congruent. Since $\angle DAC$ is right, it will follow at once that $\angle BAC$ is also a right angle.

The two triangles share a side, AC, and, by construction, $AD = AB$. By applying the Pythagorean theorem to the right-angled triangle DAC, you get

$$CD^2 = AD^2 + AC^2 = AB^2 + AC^2 = BC^2.$$

Hence $CD = BC$. But now the triangles BAC and DAC have equal sides, and hence, by Proposition I.8, they are congruent, as required. Q.E.D.

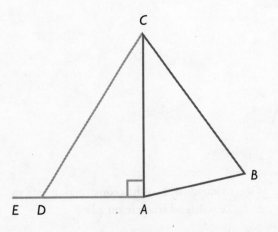

FIGURE **4.4** Proof of the converse of the Pythagorean theorem.

Book II of *Elements* deals with geometric algebra, establishing in a geometric fashion results that nowadays are generally handled algebraically, for example, the identity

$$(a + b)^2 = a^2 + 2ab + b^2.$$

Book III presents thirty-seven results about circles, including a proof that any angle inscribed in a semicircle is right. Euclid's proof is so elegant that, as with Proposition I.48, I can't resist giving it here (see Figure 4.5 for the diagram).

Proposition III.31. An angle inscribed in a semicircle is a right angle.

PROOF: Draw a semicircle with center O and diameter BC. Let A be any point on the semicircle. The theorem asserts that $\angle BAC$ is a right angle.

Draw the radius OA, and let $\angle BAO = r$, $\angle CAO = s$. Since AO and BO are radii of the semicircle, triangle ABO is isosceles. Hence, as base angles of an isosceles triangle are equal, $\angle ABO = \angle BAO = r$. Similarly, triangle AOC is isosceles, and so $\angle ACO = \angle CAO = s$.

But the angles of a triangle sum to two right angles. Applying this fact to triangle ABC,

$r + s + (r + s)$ = two right angles,

which simplifies to give

$2(r + s)$ = two right angles.

Thus $r + s$ is equal to one right angle. But $\angle BAC = r + s$, so the proof is complete. Q.E.D.

Book IV of *Elements* includes constructions of regular polygons, polygons whose sides are all equal and whose angles are all equal—the simplest examples being the equilateral triangle and the square. Book V is devoted to an exposition of Eudoxus' theory of proportions, a geomet-

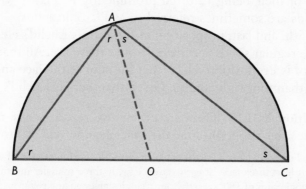

FIGURE **4.5** Proof of Euclid's Proposition III.31.

ric theory designed to circumvent the difficulties raised by the Pythagoreans' discovery that $\sqrt{2}$ is not rational. This work was largely superseded by the nineteenth-century development of the real-number system. The results in Book V are used in Book VI, in which Euclid presents a study of similar figures, two polygons being called *similar* if their angles are respectively equal and the sides about the equal angles are in proportion.

Book VI marks the end of Euclid's treatment of plane geometry. Books VII to IX are devoted to number theory, and Book X concerns measurement. Geometry becomes the focus once again in the final three books, this time the geometry of three-dimensional objects. Book XI contains thirty-nine propositions concerning the basic geometry of intersecting planes. One major result (Proposition XI.21) says that the sum of the plane angles converging at the apex of a polygonal-faced solid, such as a pyramid, is less than four right angles.

In Book XII, the study commenced in Book XI is taken further with the aid of Eudoxus' method of exhaustion. Among the results proved are that the area of a circle is proportional to the square of its diameter (Proposition XII.2).

The final book of *Elements*, Book XIII, presents eighteen propositions on the regular solids. These are three-dimensional figures having planar faces, with each face a regular polygon, all faces congruent, and all angles between pairs of adjacent faces the same. From early times, the Greeks had known of five such objects, illustrated in Figure 4.6:

- the tetrahedron, having four faces, each one an equilateral triangle
- the cube, having six square faces
- the octahedron, having eight equilateral triangles as faces
- the dodecahedron, having twelve regular pentagons as faces
- the icosahedron, having twenty equilateral triangles as faces

As a result of their being featured prominently in Plato's writings, the regular solids are sometimes referred to as Platonic solids.

The 465th, and last, proposition in *Elements* is Euclid's elegant proof that the five regular solids are the only ones there are. All that is needed for the proof is Proposition XI.21, that the sum of the face angles at any apex is less than four right angles. Given that, here is Euclid's argument.

Proposition XIII.18. There are exactly five regular solids: the tetrahedron, the cube, the octahedron, the dodecahedron, and the icosahedron.

PROOF: For a regular solid with triangular faces, the face angles are all 60°, and there must be at least three faces at each vertex. Exactly three triangles meeting at a vertex gives an angle sum of 180°; exactly four triangles gives an angle sum of 240°; and exactly five triangles an angle sum of 300°. With six or more triangles meeting at a ver-

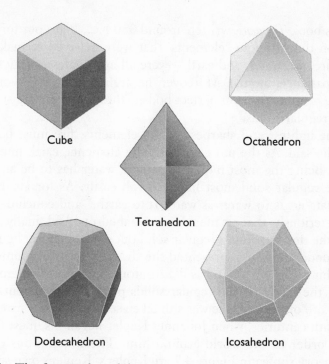

Cube

Octahedron

Tetrahedron

Dodecahedron

Icosahedron

FIGURE 4.6 The five regular solids.

tex, the angle sum would be equal to, or greater than, 360°, which is impossible by Proposition XI.21. So there are at most three regular solids with triangular faces.

For a regular solid with square faces, if three faces meet at a vertex, the angle sum at that vertex is 270°. But if four or more squares were to meet at a vertex, the angle sum would be 360° or more, so by Proposition XI.21 again, that cannot occur. Hence there can be at most one regular solid having square faces.

The interior angle at a vertex of a pentagon is 108°, so by Proposition XI.21, the only possible regular solid with pentagonal faces will have three faces meeting at each vertex, with an angle sum of 324°.

The interior angle at a vertex of any regular polygon with six or more sides is at least 120°. Since 3 × 120° = 360°, Proposition XI.21 implies that there can be no regular solid whose faces have six or more sides.

These considerations imply that the five regular solids already listed are the only ones possible. Q.E.D.

Geometry as a theory of everything

Struck by the beauty and the logical elegance of geometry, several mathematicians and philosophers have sought to use geometric ideas to explain the universe in which we live. One of the first of these was Plato, who was so enamored of the five regular solids that he used them as the basis for an early atomic theory of matter.

In his book *Timaeus*, written around 350 B.C., Plato put forward the suggestion that the four 'elements' that were believed to make up the world—fire, air, water, and earth—were all aggregates of tiny solids (in modern parlance, atoms). Moreover, he argued, since the world could only have been made from perfect bodies, these elements must have the shape of regular solids.

As the lightest and sharpest of the elements, fire must be a tetrahedron, he said. As the most stable of the elements, earth must consist of cubes. Being the most mobile and fluid, water has to be an icosahedron, the regular solid most likely to roll easily. As for air, Plato observed that "air is to water as water is to earth," and concluded, somewhat mysteriously, that air must be an octahedron. And finally, so as not to leave the one remaining regular solid out of the picture, he proposed that the dodecahedron represented the shape of the entire universe.

As whimsical and fanciful as Plato's atomic theory may seem to modern eyes, the idea that the regular solids played a fundamental role in the structure of the universe was still taken seriously in the sixteenth and seventeenth centuries, when Johannes Kepler began his quest for mathematical order in the world around him. The illustrations of Plato's atomic theory shown in Figure 4.7 are Kepler's. Kepler's own suggestion of the role played by the regular solids in the universe may strike the modern reader as somewhat more scientific than Plato's atomic theory—though it's still wrong. Here it is.

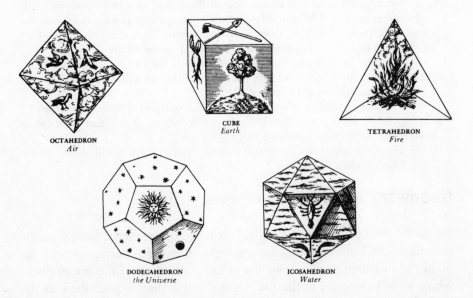

OCTAHEDRON
Air

CUBE
Earth

TETRAHEDRON
Fire

DODECAHEDRON
the Universe

ICOSAHEDRON
Water

FIGURE **4.7** Plato's atomic theory of matter, as illustrated by Johannes Kepler.

There were six known planets in Kepler's time: Mercury, Venus, Earth, Mars, Jupiter, and Saturn. Influenced by Copernicus's theory that the planets move around the sun, Kepler tried to find numeric relations to explain why there were exactly six planets, and why they were at their particular distances from the sun. He decided eventually that the key was not numeric, but geometric. There were precisely six planets, he reasoned, because the distance between each adjacent pair must be connected with a particular regular solid, of which there are just five.

After some experimentation, he found an arrangement of nested regular solids and spheres so that each of the six planets had an orbit on one of six spheres. The outer sphere (on which Saturn moves) contains an inscribed cube, and on that cube is inscribed in turn the sphere for the orbit of Jupiter. In that sphere is inscribed a tetrahedron, and Mars moves on that figure's inscribed sphere. The dodecahedron inscribed in the Mars-orbit sphere has the Earth-orbit sphere as its inscribed sphere, in which the inscribed icosahedron has the Venus-orbit sphere inscribed. Finally, the octahedron inscribed in the Venus-orbit sphere has itself an inscribed sphere, on which the orbit of Mercury lies.

To illustrate his theory, Kepler drew the painstakingly detailed picture reproduced in Figure 4.8. Clearly, he was most pleased with what he had done. The only problem is that it is all nonsense!

First, the correspondence between the nested spheres and the planetary orbits is not really exact. Having himself been largely responsible for producing accurate data on the planetary orbits, Kepler was certainly aware of the discrepancies, and tried to adjust his model by taking the spheres to be of different thicknesses, though without giving any reason why their thicknesses should differ.

Second, as we now know, there are not six, but at least nine planets. Uranus, Neptune, and Pluto were discovered subsequent to Kepler's time—a totally devastating discovery for any theory like Kepler's, based on the regular solids, of which there are only five.

From the modern standpoint, it may seem hard to believe that two intellectual giants of the caliber of Plato and Kepler should have proposed such crackpot theories. What drove them to seek connections between the regular solids and the structure of the universe?

The answer is that they were driven by the same deep-seated belief that motivates today's scientists: that the pattern and order in the world can all be described, and to some extent explained, by mathematics. At the time, Euclid's geometry was the best-developed branch of mathematics, and the theory of the regular solids occupied a supreme position within geometry; a complete classification had been achieved, with all five regular solids having been identified and extensively studied. Though

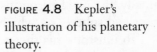

FIGURE **4.8** Kepler's illustration of his planetary theory.

ultimately not sustainable, Kepler's theory was, in its conception, extremely elegant, and very much in keeping with the view expressed by his contemporary, Galileo: "The great book of nature can be read only by those who know the language in which it was written. And this language is mathematics." Indeed, it was Kepler's fundamental belief in mathematical order that led him to adjust his mathematical model in order to fit the observed data, pursuing the aesthetic elegance of the model even at the cost of a 'fudge' that he could not explain.

In the particulars, both Plato and Kepler were most certainly off target with their atomic theories. But in seeking to understand the patterns of nature through the abstract patterns of mathematics, they were working within a tradition that continues to this day to be highly productive.

Slicing through cones

One further piece of Greek geometry should be mentioned, though since it postdates Euclid's *Elements* by several generations, it cannot strictly be considered Euclidean geometry: the work on the conic sections contained in Apollonius' eight-volume treatise *Conics*.

The conic sections are curves produced when a slice is cut through a cone, as illustrated in Figure 4.9. There are three such curves: the ellipse, the parabola, and the hyperbola. These curves had been extensively studied throughout the Greek era, but it was in *Conics* that this work was all brought together and systematized, in much the same way that *Elements* organized what was known in Euclid's time.

Along with *Elements* and the work of Archimedes, Apollonius' *Conics* was still a highly regarded textbook in the seventeenth century, when Kepler made his profound observation that the planets travel around the sun in elliptical orbits. This discovery demonstrated that, far from being purely of aesthetic interest, one of the conic sections—the ellipse—was the very path by which the planets, and humanity along with them, travel through space.

In fact, the shape of planetary orbits is not the only way in which conic sections figure in the physics of motion. When a ball or other projectile is thrown into the air, it follows a parabolic path, a fact utilized by the mathematicians who prepare artillery tables to ensure that wartime gunners can accurately aim their shells to hit the desired target.

An oft-repeated myth is that Archimedes made use of the properties of the parabola to defend the city of Syracuse against the Roman invaders in their war against Carthage. According to this story, the great Greek mathematician built huge parabolic mirrors that focused the sun's

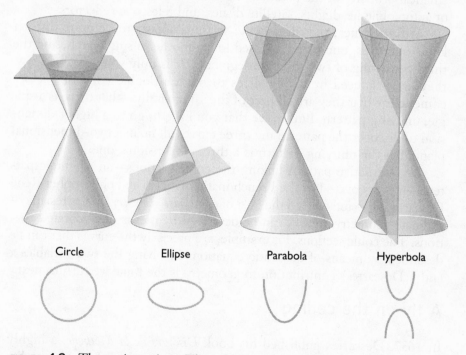

Circle   Ellipse   Parabola   Hyperbola

FIGURE **4.9** The conic sections. These four curves are obtained when a hollow double cone is sliced by a plane.

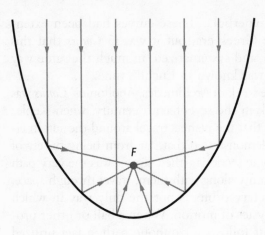

FIGURE **4.10** Rays approaching a parabola parallel to its axis are reflected to its focus (*F*). This geometric fact has found a number of applications, both in wartime and in peacetime.

rays on the enemy ships, setting them ablaze. The mathematical property involved here is that rays that fall on the parabola parallel to its axis are all reflected to a single point, known as the parabola's *focus* (see Figure 4.10). Given the difficulties that would be encountered in trying to aim such a device, this story is unlikely to be true. However, the same mathematical property of the parabola is successfully used today in the design of automobile headlights, satellite dishes, and telescope reflectors.

At first glance, the three conic sections appear to be quite distinct kinds of curves, one being a closed loop, another a single arch, and the third consisting of two separate segments. It is only when you see how they are produced by taking slices through a double cone that it becomes clear that they are all part of the same family—that there is a single, unifying pattern. But notice that you have to go to a higher dimension to discover the pattern: the three curves all lie in a two-dimensional plane, but the unifying pattern is a three-dimensional one.

An alternative pattern linking the conic sections—an algebraic pattern—was discovered by the French mathematician and philosopher René Descartes, who introduced the idea of *coordinate geometry*. In Cartesian coordinate geometry, figures can be described by means of algebraic equations. The conic sections, for example, are precisely the curves that can be described by means of quadratic equations involving the two variables x and y. Descartes's contribution to geometry is the topic we turn to next.

A fly on the ceiling

In 1637, Descartes published his book *Discours de la Méthode*, a highly original philosophical analysis of the scientific method. In an appendix, entitled *La Géomètrie*, he presented the mathematical world with a revolutionary new way to do geometry: with algebra. Indeed, the revolution that Descartes set in motion with this publication was to be a complete one, in that his new approach not only enabled mathematicians to make use of algebraic techniques in order to solve problems in geometry but also effectively gave them the option of regarding geometry as a branch of algebra.

Descartes's key idea was (taking the two-dimensional case) to introduce a pair of coordinate axes: two real-number lines drawn at right angles, as shown in Figure 4.11. The point of intersection of the two axes is called the origin (of the coordinates). The two axes are most frequently labeled the x-axis and the y-axis; the origin is generally denoted by 0.

In terms of the coordinate axes, every point in a plane has a unique name as a pair of real numbers: the x-coordinate and the y-coordinate of the point. The idea, then, is to represent geometric figures with algebraic expressions involving x and y; in particular, lines and curves are represented by algebraic equations involving x and y.

For example, a straight line with slope m that crosses the y-axis at the point $y = c$ has the equation

$$y = mx + c.$$

The equation of a circle with its center at the point (p, q) and radius r is

$$(x - p)^2 + (y - q)^2 = r^2.$$

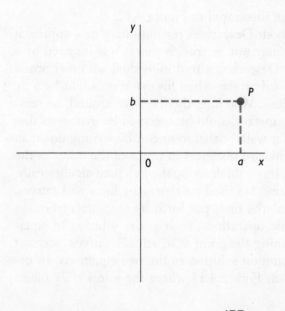

FIGURE **4.11** The coordinate axes introduced by Descartes. Relative to a given pair of coordinate axes, every point in a plane has a unique name as a pair of real numbers. The point P shown has coordinates (a,b), where a,b are real numbers.

Expanding the two bracketed terms in this equation and rearranging, the expression becomes

$$x^2 + y^2 - 2px - 2qy = k,$$

where $k = r^2 - p^2 - q^2$. In general, any equation of this last form for which $k + p^2 + q^2$ is positive will represent a circle with center (p, q) and radius $\sqrt{[k + p^2 + q^2]}$. In particular, the equation of a circle of radius r whose center is at the origin is

$$x^2 + y^2 = r^2.$$

An ellipse centered about the origin has an equation of the form

$$\frac{x^2}{a^2} + \frac{y^2}{b^2} = 1.$$

A parabola has an equation of the form

$$y = ax^2 + bx + c.$$

Finally, a hyperbola centered about the origin has an equation of the form

$$\frac{x^2}{a^2} - \frac{y^2}{b^2} = 1,$$

or (a special case) of the form

$$xy = k.$$

Examples of these curves are illustrated in Figure 4.12.

An oft-repeated story about Descartes's revolutionary new approach to geometry, which may or may not be true, is that it was inspired by a fly. According to the story, Descartes, a frail individual who was prone to sickness, was lying ill in bed on day when his eye was caught by a fly crawling around on the ceiling. Watching the fly move around, he realized that its position at any instant could be specified by giving its distance from two perpendicular walls at that moment. By writing down an equation that gives one of the two distances in terms of the other as the fly moves across the ceiling, he could describe the fly's path algebraically.

When algebraic equations are used to represent lines and curves, geometric arguments such as the ones put forth by the ancient Greeks may be replaced by algebraic operations, such as the solution of equations. For instance, determining the point at which two curves intersect corresponds to finding a common solution to the two equations. In order to find the points P, Q in Figure 4.13 where the line $y = 2x$ meets

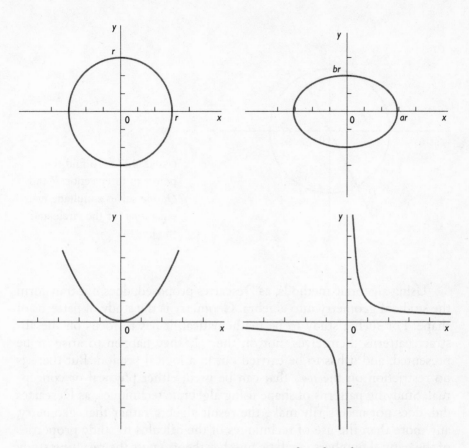

FIGURE **4.12** Graphs of conics. *Top left:* The circle $x^2 + y^2 = r^2$. *Top right:* The ellipse $x^2/a^2 + y^2/b^2 = r^2$. *Bottom left:* The parabola $y = x^2$. *Bottom right:* The hyperbola $xy = 1$.

the circle $x^2 + y^2 = 1$, you solve these two equations simultaneously. Substituting $y = 2x$ in the second equation gives $x^2 + 4x^2 = 1$, which solves to give $x = \pm 1/\sqrt{5}$. Using the equation $y = 2x$ to obtain the corresponding y-coordinates of the points of intersection, you get $P = (1/\sqrt{5}, 2/\sqrt{5})$, $Q = (-1/\sqrt{5}, -2/\sqrt{5})$.

To give another example, the perpendicularity of two straight lines corresponds to a simple algebraic condition on their equations: the lines $y = mx + c$ and $y = nx + d$ are perpendicular if, and only if, $mn = -1$.

By incorporating methods of the differential calculus, questions to do with lines tangent to curves may also be dealt with algebraically. For instance, the line tangent to the curve $y = f(x)$ at the point $x = a$ has the equation

$$y = f'(a)x + [f(a) - f'(a)a].$$

157

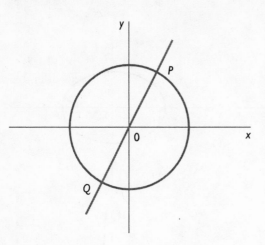

FIGURE **4.13** To find the points of intersection P and Q, one solves simultaneous equations for the circle and the line.

Using algebraic methods, as Descartes proposed, does not transform the study of geometry into algebra. Geometry is the study of patterns of shape. For such a study to be mathematical, it has to focus on the abstract patterns themselves, not on the way they happen to arise or be presented, and it has to be carried out in a logical fashion. But there is no restriction on the *tools* that can be used, either physical or conceptual. Studying patterns of shape using algebraic techniques, as Descartes did, does not necessarily make the result algebra rather than geometry, any more than the use of techniques of the calculus to study properties of the natural numbers (analytic number theory) puts the resulting study outside the boundaries of number theory.

The use of algebraic techniques (and calculus-based analytic techniques) in geometry provides a degree of precision, and a potential for greater abstraction, that takes the study of shape into realms that would otherwise remain forever inaccessible. One early illustration of the power of these techniques was the resolution, toward the end of the nineteenth century, of three geometric problems that had defied solution since the time of the Greeks.

Squaring the circle and other impossibilities

Computing the area of a square or rectangle whose dimensions are known is an easy matter involving nothing more complicated than multiplication. Computing the area of a figure with curved edges, such as a circle or an ellipse, is far more difficult. The Greeks used the method of exhaustion; present-day mathematicians use the integral calculus. Both techniques are considerably more complicated than multiplication.

Another possible method is to find a square whose area is the same as the curved figure, and then compute the area of that square in the normal fashion. Can such a square be found, and if so, how? This is the problem of *quadrature* of a given figure, a problem the Greeks spent considerable time trying to solve. The simplest case—at least to state—is the problem of quadrature of the circle: given a circle, find a square having the same area.

Not surprisingly, the Greeks wondered if there were a construction using only ruler and compass, the 'pure' tools favored by Euclid in *Elements*, but they were unable to produce one. Nor did countless succeeding generations of geometers fare any better, either the professional mathematicians or any of the many amateur would-be 'circle-squarers' who tried their hands at the problem. (It should be noted that the problem asks for an exact solution; there are a number of ruler-and-compass methods for obtaining approximate answers.)

In 1882, the German mathematician Ferdinand Lindemann finally brought the quest to an end by proving conclusively that ruler-and-compass quadrature of the circle is impossible. His proof was purely algebraic, by way of Cartesian coordinates, and goes like this.

First of all, operations performed with ruler and compass have corresponding algebraic descriptions. Fairly elementary considerations show that any length that can be constructed using ruler and compass can be computed, starting from integers, using only a sequence of additions, multiplications, subtractions, divisions, and the extraction of square roots. It follows that any length that can be constructed using the classical Greek tools must be what is nowadays called *algebraic*; that is, it can be obtained as the solution to a polynomial equation of the form

$$a_n x^n + a_{n-1} x^{n-1} + \ldots + a_2 x^2 + a_1 x + a_0 = 0,$$

where the coefficients $a_n, \ldots, a_0$ are all integers.

Real numbers that are not algebraic are called *transcendental*. Transcendental numbers are all irrational, but the converse is not true; for instance, $\sqrt{2}$ is irrational, but it is certainly not transcendental, being a solution to the equation

$$x^2 - 2 = 0.$$

What Lindemann did was to prove that the number π is transcendental. His proof used methods of the calculus.

That π is transcendental at once implies that quadrature of the circle is impossible. Here's why. Consider the unit circle, the circle with radius $r = 1$. The area of the unit circle is $\pi \times r^2 = \pi \times 1 = \pi$. So, if you could construct a square having the same area as the unit circle, the square would have area π, and its edges would have length $\sqrt{\pi}$. In this way, therefore, you would be able to construct a length of $\sqrt{\pi}$ using ruler and compass. This would imply that $\sqrt{\pi}$ was algebraic, and it would follow at once that π was algebraic, contrary to Lindemann's result.

Another classical Greek problem that was resolved with the aid of Cartesian techniques was that of duplicating the cube: given any cube, find the side of another cube having exactly twice the volume. Once again, the problem, as posed by the Greeks, was to find a construction using ruler and compass only, and once again, the problem can be solved if this restriction is relaxed. Duplication of the cube is easily accomplished by means of a so-called *neusis construction*, which is performed using a compass and a *marked* ruler that may be slid along some configuration of lines until a certain condition is met. Another solution uses conic sections, and there is a three-dimensional construction involving a cylinder, a cone, and a torus.

Here is the solution. If you can duplicate the unit cube, the cube with a volume of 1, then the duplicate has a volume of 2, and hence its edges must have a length of $\sqrt[3]{2}$. In algebraic terms, therefore, duplication of the unit cube corresponds to solving the cubic equation $x^3 = 2$. An argument somewhat simpler than Lindemann's shows that this equation cannot be solved by means of a sequence of the basic arithmetical operations (addition, multiplication, subtraction, and division) and the extraction of square roots, the operations that correspond to ruler-and-compass constructions. Hence the unit cube cannot be duplicated using only ruler and compass.

A third Greek problem that achieved some fame as a result of defying solution for many years was that of trisecting an angle. The problem is to find a method, using ruler and compass only, that, for any given angle, will produce an angle exactly one-third its size.

The problem can be solved immediately for some angles. For example, trisecting a right angle is easy: you simply construct a 30° angle. But the problem asks for a method that works in all cases. Again, it is possible to solve the general problem if the restriction to ruler and compass is relaxed; in particular, trisection of any angle is easy by means of a neusis construction.

In algebraic terms, trisection of an arbitrary angle amounts to the solution of a cubic equation, which, as we just observed, is not possible using only the basic arithmetical operations together with the extraction of square roots. Thus, trisection cannot be achieved using ruler and compass alone.

It should be re-emphasized that the above three problems were not, in themselves, major mathematical problems. Restriction to ruler-and-compass constructions was largely a Greek intellectual game. The Greeks could solve all three problems when this restriction was removed. It was the fact that the problems as stated defied solution for so long that led to their fame.

It is interesting that, in each case, the solution came only after the problem was translated from a purely geometric task to an algebraic one, a transformation that enabled other techniques to be brought to bear. As originally formulated, the three problems concerned patterns (i.e., sequences) of geometric construction using certain tools; the solutions depended on reformulating the problems in terms of equivalent algebraic patterns.

The startling discovery of non-Euclidean geometries

As mentioned on page 144, from the first formulation of Euclid's postulates for geometry, the fifth, or 'parallel postulate', was regarded as problematic. Its truth was never questioned—everyone agreed it was 'obvious'. But mathematicians felt that it was not sufficiently fundamental to be taken as an axiom; rather, it ought to be proved as a theorem.

The obviousness of the axiom seemed to be emphasized by the various alternative formulations that were obtained. In addition to Euclid's original statement—one of the least obvious formulations, as it happens—each of the following can be shown to be completely equivalent to the fifth postulate:

Playfair's postulate: Given a straight line and a point not on the line, exactly one straight line may be drawn through the point, parallel to the given line. (The formal definition of 'parallel' is that two lines are parallel if, however far they be extended, they do not meet.)

Proclus' axiom: If a line intersects one of two parallel lines, it must intersect the other also.

Equidistance postulate: Parallel lines are everywhere equidistant.

Triangle postulate: The sum of the angles of a triangle is two right angles.

Triangle area property: The area of a triangle can be as large as we please.

Three points property: Three points either lie on a line or lie on a circle.

Most people find one or more of these statements 'obvious', generally the first three in the list, and maybe also the triangle area property. But why? Take Playfair's postulate as an example. How do you know it is true? How could you test it?

Suppose you draw a line on a sheet of paper and mark a point not on the line. You are now faced with the task of showing that there is one, and only one, line parallel to the given line that passes through the chosen point. But there are obvious difficulties here. For one, no matter how fine the point of your pencil, the lines you draw still have a definite thickness, so how do you know where the actual *lines* are? Second, in order to check that your second line is in fact parallel to the first, you would have to extend both lines indefinitely, which is not possible. (You can draw many lines through the given point that do not meet the given line *on the paper*.)

Thus, Playfair's postulate is not really suitable for experimental verification. How about the triangle postulate? Certainly, verifying this postulate does not require extending lines indefinitely; it can all be done 'on the page'. Admittedly, it is likely that no one has any strong intuition concerning the angle sum of a triangle being 180°, the way we do about the existence of unique parallels, but since the two statements are entirely equivalent, the absence of any supporting intuition does not affect the validity of the triangle approach.

Suppose, then, that you could draw triangles and measure their three angles with sufficient accuracy that you could calculate the sum of the angles to within 0.001 of a degree, a perfectly reasonable assumption in this day and age. You draw a triangle and measure the angles. If these add up to 180°, then all you can conclude for certain is that the angle sum is between 179.999° and 180.001°, which is inconclusive.

On the other hand, it is, in principle, possible to demonstrate the falsity of the fifth postulate in this way. If you found a triangle whose angle sum worked out to be 179.9°, then you would know for certain that the angle sum was between 179.899° and 179.901°, and so the answer could not possibly be 180°.

According to mathematical folklore, Gauss himself made an attempt to check the fifth postulate experimentally in the early part of the nineteenth century. To avoid the problem of drawing infinitely thin straight lines, he took light to be the straight lines, and to minimize the effect of errors in measurement, he worked with a very large triangle, whose apexes were located on three mountaintops. Setting up a fire on one of them, and using mirrors to reflect the light, he created a huge triangle of light rays. The angle sum worked out to be 180° plus or minus the experimental error. With the greatest optimism in the world, that experimental error could be no less than 30 angular seconds. So the experiment could have proved nothing, except that on a scale of several miles, the angle sum of a triangle is *fairly close to* 180°.

In fact, there seems to be no sound basis in our everyday experience to support the fifth postulate. And yet, in a form such as Playfair's postulate, we believe it—indeed, we find it obvious. The abstract concept of a straight line, having length but no width, seems perfectly sound, and we can in fact visualize such entities; the idea of two such lines being indefinitely extended, and remaining everywhere equidistant and not meeting, seems meaningful; and we have a deep-rooted sense that parallels exist and are unique.

As I hinted at the very beginning of this chapter, these fundamental geometric notions, and the intuitions that accompany them, are not part of the physical world we live in; they are part of ourselves, of the way we are constructed as cognitive entities. Euclidean geometry may or may not be the way the world is 'made up', whatever that may mean, but it does appear to capture the way human beings *perceive* the world.

So where does all of this leave the geometer? On what does she base her subject, a subject that deals with 'points', 'lines', 'curves', and the like, all of which are abstract idealizations, molded by our perceptions? The answer is that, when it comes to establishing the theorems that represent mathematical truths, the axioms are, quite literally, all there is. Practical experience and physical measurement cannot give us the certainty of mathematical knowledge. In constructing a proof in geometry, we may well rely upon mental pictures of straight lines, circles, and so forth in order to guide our reasoning process, but our proof must rest entirely upon what the axioms tell us about those entities.

Despite Euclid's attempt to axiomatize geometry, it took mathematicians two thousand years to come to terms fully with the significance of this last remark, and to abandon the intuition that told them that Euclidean geometry was self-evidently the geometry of the universe we live in.

The first significant step toward this realization was taken by an Italian mathematician named Girolamo Saccheri. In 1733, he published a two-volume work titled *Euclid Freed of Every Flaw*. In this book, he tried to establish the fifth postulate by showing that its negation resulted in a contradiction.

Given a line and a point not on the line, there are three possibilities for the number of parallel lines through that point:

1. there is exactly one parallel;
2. there are no parallels;
3. there is more than one parallel.

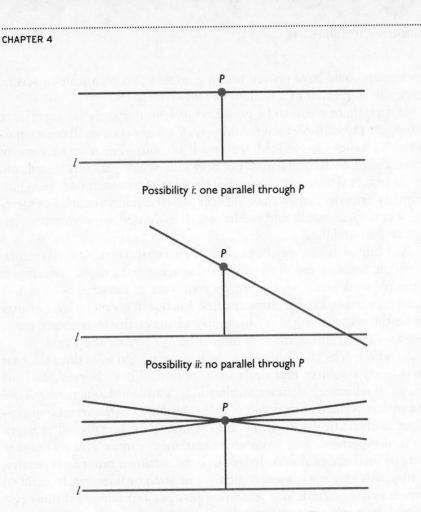

Possibility *i*: one parallel through *P*

Possibility *ii*: no parallel through *P*

Possibility *iii*: many parallels through *P*

FIGURE **4.14** The parallel postulate. Given a line *l* and a point *P* not on *l*, there are three possibilities for the existence of lines through *P* that are parallel to *l*: exactly one parallel, no parallels, or more than one parallel. The parallel postulate says that the first possibility occurs.

These possibilities are illustrated in Figure 4.14.

Possibility 1 is Euclid's fifth postulate. Saccheri set out to demonstrate that each of the other two possibilities leads to a contradiction. Assuming that Euclid's second postulate requires straight lines to be infinitely long, he found that possibility 2 leads to a contradiction. He was less successful in eliminating possibility 3. He obtained a number of consequences of statement 3 that were counterintuitive, but was unable to derive a formal contradiction.

A hundred years later, four different mathematicians, all working independently, tried the same approach. But they took a crucial additional step that their predecessor had not: they managed to break free of the belief that there is only one geometry, the one in which the fifth postulate holds.

The first was Gauss. Working with a formulation equivalent to possibility 3, that the angle sum of any triangle is less than two right angles, he came to realize that possibility 3 probably does not lead to an inconsistency, but rather to a strange alternative geometry—a *non-Euclidean* geometry. Exactly when Gauss did this work is not known, since he did not publish it. The first reference we have is in a private letter written to a colleague, Franz Taurinus, in 1824, in which he says,

> The assumption that the sum of the three angles is less than 180° leads to a curious geometry, quite different from ours, but thoroughly consistent, which I have developed to my entire satisfaction.

In another letter, written in 1829, he makes it clear that his reason for withholding his findings from publication is his fear that it would harm his considerable reputation if he were to go on record as saying that Euclidean geometry is not the only one possible.

Such are the pressures on the most successful. No such constraints hampered János Bolyai, a young Hungarian artillery officer whose father, a friend of Gauss, had himself worked on the parallel axiom. Though the senior Bolyai advised his son not to waste his time on the problem, János took no notice, and was able to take the bold step his father had not. As Gauss had done, he recognized that possibility 3 leads not to an inconsistency, but to a completely new geometry. It was not until János's work was published as an appendix to a book of his father's in 1832 that Gauss informed the two men of his own earlier observations on the subject.

Sadly for Bolyai, not only had Gauss beaten him to the idea, and not only was he never widely recognized for his work during his lifetime, but it turned out that he was not even the first to publish on the new non-Euclidean geometry. Three years previously, in 1829, Nikolay Lobachevsky, a lecturer at Kazan University in Russia, had published much the same results under the title (translated from Russian) *Imaginary Geometry*.

So now there were two geometries, Euclidean geometry, with its unique parallels, and another geometry with multiple parallels, known nowadays as *hyperbolic geometry*. It was not long before a third was added. In 1854, Bernhard Riemann reexamined the contradiction that Saccheri had derived from possibility 2—the statement that there are no parallels to a given line through a given point—and observed that this contradiction could be avoided. The crucial mistake Saccheri had made was to assume that Euclid's second postulate (a finite straight line can be produced continuously) implies the infinitude of the line. This assumption is simply not valid.

Riemann suggested that possibility 2 may be consistent with Euclid's first four postulates, and that, if so, then when all five postulates are taken together as axioms, the result is yet another geometry, *Riemannian geometry*, in which the angles of a triangle always add up to more than two right angles.

None of the actors in this drama suggested that Euclidean geometry was not the right one for the universe we live in. They simply advocated the view that Euclid's first four postulates do not decide between the three possibilities for the parallel postulate, each one of which leads to a consistent geometry. Moreover, none of them provided a *proof* that possibilities 2 or 3 were consistent with Euclid's other postulates; rather, they formed the *opinion* that this was so, based on their work with the postulate concerned. Proof that the two non-Euclidean geometries are consistent (more precisely, that they are as consistent as Euclidean geometry itself) came in 1868, when Eugenio Beltrami showed how to interpret the hyperbolic and Riemannian geometries within Euclidean geometry.

It is at this point that we must come to grips with what it means to abandon one's intuition and work in a purely axiomatic fashion.

The basic undefined objects in Euclidean geometry are points and straight lines. (Circles may be defined: a circle is the collection of all points equidistant from a given point.) We may well have various mental pictures of and intuitions about these objects, but unless these pictures or intuitions are captured by our axioms, they are logically irrelevant as far as the geometry is concerned.

Euclidean geometry works for a plane. Consider the geometry that holds on the surface of the earth, which we shall assume to be a perfect sphere; call this geometry 'spherical geometry'. Though the earth is a three-dimensional object, its surface is two-dimensional, so its geometry will be a two-dimensional geometry. What are the 'straight lines' in this geometry? The most reasonable answer is that they are the surface *geodesics*; that is, the straight line from point A to point B is the path of shortest distance from A to B on the surface. For a sphere, the straight line from A to B is the great circle route from A to B, illustrated in Figure 4.15. Viewed from above the earth, such a path does not appear to be a straight line, since it follows the curve of the earth's surface. But as far as the geometry of the surface itself is concerned, it appears to have all the properties of a straight line. For example, an airplane flying the shortest route from New York (point A in Figure 4.15) to London (point B) would follow such a path.

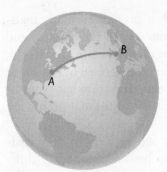

FIGURE **4.15** A straight line in spherical geometry is the shortest distance from point *A* to point *B*, namely, the great circle route.

Spherical geometry satisfies the first four of Euclid's postulates, though in the case of the second postulate, notice that the continuous extendibility of a straight line does not imply infinitude. Rather, when a straight line segment is extended, it eventually wraps around the globe and meets up with itself. The fifth postulate is not satisfied. Indeed, in this geometry there are no parallels; any two straight lines meet at two antipodal points, as shown in Figure 4.16.

Thus, the axioms of Riemannian geometry are true of the geometry of the surface of a sphere—at least, they are true given the way Euclid stated his postulates. However, it is generally assumed that Euclid intended his first postulate to imply that two points determine exactly one straight line, and the same should be true of Riemannian geometry. Spherical geometry does not have this property; any pair of antipodal points lie on infinitely many straight lines, namely, every great circle through those two points. To obtain a world in which this strengthening of the first postulate is valid, we can declare that antipodal points are one and the same point. Thus, the North Pole is declared to be exactly the same point as the South Pole, any point on the equator is identified with the diametrically opposite point on the equator, and so on. Of course, this process creates a geometric monster that defies visualization.

FIGURE **4.16** There are no parallels in spherical geometry; any two straight lines meet in a pair of antipodal points.

But for all that you cannot picture this world, it provides a mathematically sound interpretation of Riemannian geometry: all the axioms, and hence all the theorems, of Riemannian geometry are true in this world, where the 'points' of the geometry are points on the surface of the 'what was a sphere' and the 'straight lines' of the geometry are what you obtain from the great circles on the sphere when the antipodal points are declared identical.

Though the Riemannian world just described appears to defy visualization, working with its geometry is not so bad in practice. A difficulty arises only when you try to think of the entire world at once. The strange identification of antipodal points is only necessary to avoid problems that arise with 'infinitely long' lines, that is, lines that stretch halfway around the globe. On a smaller scale, there is no such problem, and the geometry is the same as spherical geometry, a geometry that globe-trotting inhabitants of Planet Earth are very familiar with.

In particular, our everyday knowledge of global travel enables us to observe, as an empirical fact, the theorem of Riemannian geometry that the angles of a triangle add up to more than two right angles. In fact, the larger the triangle, the greater the angle sum. For an extreme case, imagine a triangle that has one apex at the North Pole and the other two apexes on the equator, one on the Greenwich meridian and the other at 90° west, as shown in Figure 4.17. Each angle of this triangle is 90°, so the angle sum is 270°. (Notice that this is a triangle in the sense of spherical geometry; each side is a segment of a great circle, and hence the figure is bounded by three straight lines in the geometry.)

The smaller the triangle, the smaller the angle sum. Triangles that occupy a very small area of the globe have angle sums very close to 180°. Indeed, for human beings living on the surface of the earth, a triangle marked out on the surface will appear to have an angle sum of exactly 180°, as in the Euclidean geometry that works in a plane, because the

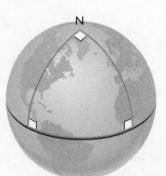

FIGURE **4.17** In spherical geometry, the angle sum of a triangle is greater than 180°.

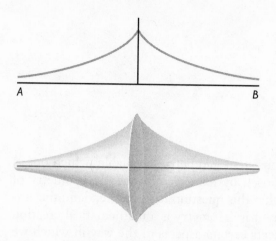

A   B

FIGURE **4.18** A double tractrix (*top*) and a pseudosphere
(*bottom*), the surface generated by rotating the tractrix
around the line *AB*

earth's curvature is effectively zero on such a small scale. But, mathematically speaking, in spherical geometry, and likewise in Riemannian geometry, the angle sum of a triangle is never equal to 180°, it is always greater.

Turning now to hyperbolic geometry, the general idea is the same—
to take the geometry of an appropriate surface, with the geodesics as the
straight lines. The question is, what surface will produce hyperbolic
geometry? The answer turns out to involve a pattern that is familiar to
every parent.

Watch a child walking along, pulling a toy attached to a string. If
the child makes an abrupt left turn, the toy will trail behind, not making a sharp corner, but curving around until it is almost behind the child
once again. This curve is called a *tractrix*.

Now take two opposite copies of a tractrix, as shown on the top in
Figure 4.18, and rotate this double curve about the line *AB*. The result-
ing surface, shown on the bottom in Figure 4.18, is called a *pseudosphere*.
It extends out to infinity in both directions.

The geodesic geometry on the pseudosphere is hyperbolic geometry. In particular, Euclid's first four postulates are true for this geometry,
as are the other Euclidean axioms that Hilbert wrote down. But the fifth
postulate is false: on a pseudosphere, given any line and any point not
on that line, there are infinitely many lines through that point that are
parallel to the given line—lines that will never meet, no matter how far
they are extended.

The angle sum of any triangle drawn on a pseudosphere is less than
180°, as illustrated in Figure 4.19. For very small triangles, in which the
curvature of the pseudosphere does not make much difference, the angle sum is close to 180°. But if you take a triangle and then start to en-
large it, the angle sum becomes less and less.

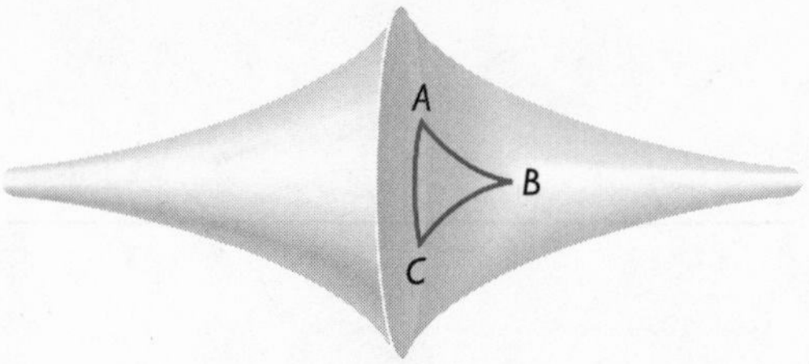

FIGURE **4.19** On a pseudo-sphere, the angle sum of a triangle is less than 180°.

So, with three equally consistent geometries available, which one is the correct one—the one chosen by nature? What is the geometry of our universe? It is not clear that this question has a single, definitive answer. The universe is the way it is; geometry is a mathematical creation of the human mind that reflects certain aspects of the way in which we encounter our environment. Why should the universe have a geometry at all?

Let's try to rephrase the question. Given that mathematics provides human beings with a very powerful means to describe and understand aspects of the universe, which of the three geometries is best suited to that task? Which geometry most closely corresponds to the observable data?

On a small, human scale—a scale that covers part or all of the earth's surface—Newtonian physics provides a theoretical framework in complete accord with the observable (and measurable) evidence, and any one of the three geometries will do. Since Euclidean geometry seems to accord with our own intuitions about the way we perceive the world, we may as well take that to be 'the geometry of the physical world'.

On the other hand, at a larger scale—from the solar system up to galaxies and beyond—Einstein's relativistic physics provides a closer fit with the observable data than does Newton's framework. At this scale, non-Euclidean geometry appears to be more appropriate. According to relativity theory, space–time is curved, the curvature manifesting itself in what we refer to as the force of gravity. The curvature of space–time is observed in the behavior of light rays, the 'straight lines' of the physical universe. When light rays from a distant star pass close to a large mass, such as the sun, their path is bent round, just as the geodesics on a globe curve around its surface.

Just which non-Euclidean geometry one uses depends on one's choice of a theory of the universe. If you assume that the present expansion of the universe will come to a halt, to be replaced by a contraction, Riemannian geometry is the most appropriate. If, on the other hand, you take the view that the universe will expand forever, then hyperbolic geometry is more suitable.

It is particularly fascinating that Einstein's relativity theory, and the astronomical observations that demonstrated its superiority over the Newtonian theory, came more than a half century after the development of non-Euclidean geometry. Here we have an example of how mathematics can move ahead of our understanding of the world. The initial abstraction of geometric patterns from observations of the world around them led the Greeks to the development of a rich mathematical theory, Euclidean geometry. During the nineteenth century, purely mathematical questions about that theory, questions concerning axioms and proofs, led to the discovery of other geometric theories. Though these alternative geometries were initially developed purely as abstract, axiomatic theories, seemingly having no applications in the real world, it turned out that they were in fact more appropriate for the study of the universe on a large scale than was Euclidean geometry.

The geometry of the Renaissance artists

For the surveyor mapping out the land or the carpenter setting out to build a house, Euclidean geometry captures the relevant patterns of shape. For the sailor or the airline pilot circumnavigating the globe, spherical geometry is the appropriate framework. For the astronomer, the geometric patterns that arise may be those of Riemannian or hyperbolic geometry. It is all a question of what you want to do and how you want to do it.

The Renaissance artists Leonardo da Vinci and Albrecht Dürer wanted to portray depth on a two-dimensional canvas. Prior to Leonardo and Dürer, it presumably never occurred to artists that there might be a means of rendering depth in their paintings. The painting shown in Plate 2 is typical of paintings before the fifteenth century in that it gives no sense of three dimensions. The painting in Plate 3 is typical of the post-Renaissance era, when artists had discovered how to create a sense of depth in a painting.

The key idea exploited by Leonardo and Dürer is to consider the surface of a painting a window through which the artist views the object to be painted. The lines of vision from the object that converge on the eye pass through that window, and the points where the lines intersect the window surface form the *projection* of the object onto that surface. The painting captures that projection, as illustrated by Dürer in the drawing shown in Figure 4.20. For the artist, therefore, the relevant patterns are those of perspective and of plane projections; the geometry that arises from these considerations is called *projective geometry*.

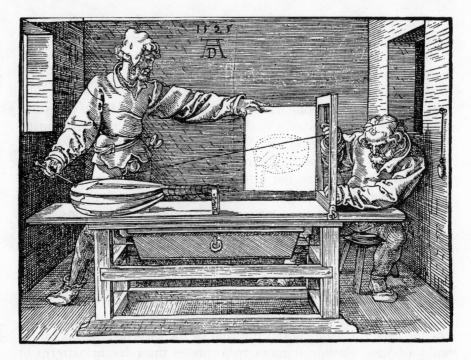

FIGURE **4.20** Albrecht Dürer's woodcut *Institutiones Geometricae* illustrates how a perspective drawing is produced by means of a projection. The glass plate held by the man on the left bears a perspective drawing of the object on the table. To create this drawing, the plate is placed in the frame held by the artist, on the right. Where the light rays (straight lines) from the object to the artist's eye intersect the plate, they determine an image of the object, known as its projection, onto the plate.

Though the fundamental ideas of perspective were discovered in the fifteenth century, and gradually came to pervade the world of painting, it was not until the end of the eighteenth century that projective geometry was studied as a mathematical discipline. In 1813, while a prisoner of war in Russia, Jean-Victor Poncelet, a graduate of the École Polytechnique in Paris, wrote the first book on the subject, *Traité des Propriètès Projectives des Figures*. During the early nineteenth century, projective geometry grew into a major area of mathematical research.

If Euclidean geometry corresponds to our mental conception of the world around us, projective geometry captures some of the patterns that enable us to see that world the way we do, for our entire visual input consists of flat, two-dimensional images on our retinas. When an artist creates a painting that is correctly in perspective, we are able to interpret it as a three-dimensional scene, as, for example, the Renaissance painting in Plate 3.

The basic idea of projective geometry is to study those figures, and those properties of figures, that are left unchanged by projection. For example, points project to points and straight lines project to straight lines, so points and lines figure in projective geometry.

Actually, you have to be a bit more careful than that, since there are two kinds of projection. One is projection from a single point, also known as central projection, which is illustrated in Figure 4.21. The other is parallel projection, sometimes referred to as projection from infinity, which is illustrated in Figure 4.22. In both illustrations, the point P is projected to the point P' and the straight line l is projected to the straight line l'. In the axiomatic study of projective geometry, however, the distinction between these two kinds of projection disappears, since in the formal theory there are no parallel lines—any two lines meet, and what were once thought of as parallel lines become lines that meet at a 'point at infinity'.

Clearly, projection distorts lengths and angles, in a way that depends on the relative positions of the objects depicted. Thus projective geometry cannot involve axioms or theorems about length, angle, or congruence. In particular, though the notion of a triangle makes sense in projective geometry, the notions of isosceles and equilateral triangles do not.

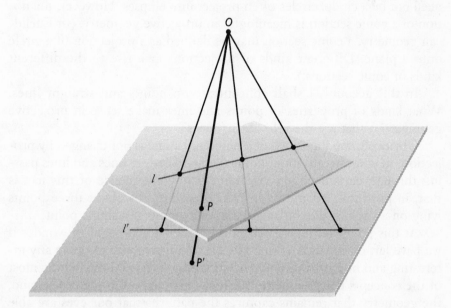

FIGURE **4.21** Projection from a single point.

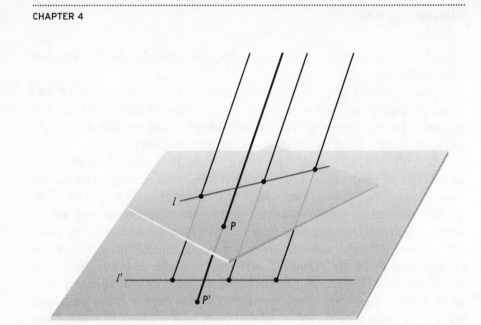

FIGURE **4.22** Projection from infinity.

The projection of any curve, however, is another curve. This fact raises the interesting question of what classifications of curves are meaningful in projective geometry. For example, the notion of a circle is obviously not one of projective geometry, since the projection of a circle need not be a circle: circles often project into ellipses. However, the notion of a conic section is meaningful in projective geometry. (In Euclidean geometry, a conic section may be defined as a projection of a circle onto a plane. Different kinds of projections give rise to the different kinds of conic sections.)

In this account, I shall concentrate on points and straight lines. What kinds of properties of points and lines make sense in projective geometry? What are the relevant patterns?

Obviously, the incidence of a point and a line is not changed by projection, so you can talk about points lying on certain lines and lines passing through certain points. An immediate consequence of this fact is that, in projective geometry, it makes sense to talk about three points lying on a straight line or of three lines meeting at a single point.

At this point, however, a skeptical reader might begin to wonder if we have left ourselves anything like enough machinery to prove any interesting and nontrivial theorems. Certainly, we have thrown away most of the concepts that dominate Euclidean geometry. On the other hand, the geometry that remains captures the patterns that our eyes are able to interpret as perspective, so projective geometry cannot be without content. The issue is, does this content manifest itself in the form of interesting geometric theorems?

Well, although there can't be any theorems involving length in projective geometry, there is a particular concept of relative length that is meaningful. It is called *cross-ratio*, and it relates any four points on a straight line.

Given three points on a line, A, B, C, a projection that sends A, B, C to three collinear points A', B', C' will, in general, change the distances AB and BC. It can also change the value of the ratio AB/BC. In fact, given any two triples of collinear points A, B, C and A', B', C', it is possible to make two successive projections that send A to A', B to B', and C to C', so you can make the ratio $A'B'/B'C'$ whatever you like.

However, in the case of four points on a line, there is a certain quantity, called the cross-ratio of the four points, that always retains its value under projections. Referring to Figure 4.23, the cross-ratio of the four points A, B, C, D is defined by

$$\frac{CA/CB}{DA/DB}.$$

Though length itself is not a notion of projective geometry, the cross-ratio is a projective notion that, at least in the form just described, is based on length. The cross-ratio figures in a number of advanced results of projective geometry.

A further indication that projective geometry is by no means impoverished is provided by a striking result proved in the early seventeenth century by the French mathematician Gérard Desargues:

> **Desargues' Theorem**. If in a plane two triangles ABC and $A'B'C'$ are situated so that the straight lines joining corresponding vertices all meet at a point O, then the corresponding sides, if extended, will intersect at three points that lie on the same line.

Figure 4.24 illustrates Desargues' theorem. If you doubt the result, draw a few further figures until you are convinced. This theorem is clearly a result of projective geometry, since it speaks only of points, straight

FIGURE **4.23** The cross-ratio of the four points A, B, C, D on a straight line is the quotient of CA/CB by DA/DB.

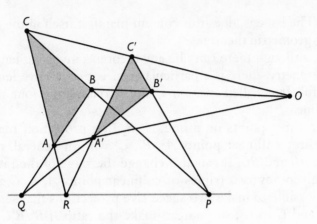

FIGURE **4.24** Desargues' theorem in the plane.

lines, triangles, lines meeting at a point, and points lying on a line, all concepts of projective geometry.

It is possible to prove Desargues' theorem using techniques of Cartesian (Euclidean) geometry, but it is by no means easy. The easiest proof is the one within projective geometry itself. Here it is.

As is the case with any theorem of projective geometry, if you can prove it for one particular configuration, it will follow for any projection of that configuration. The key step, which, on the face of it, seems much more difficult, is to try to prove the three-dimensional version of the theorem, in which the two given triangles lie in different, nonparallel planes. This more general version is illustrated in Figure 4.25. The original two-dimensional theorem is, quite clearly, a projection of this three-dimensional version onto a single plane, so proving the more general version will at once yield the original theorem.

Referring to the diagram, notice that AB lies in the same plane as $A'B'$. In projective geometry, any two lines in the same plane must meet. Let Q be the point where AB and $A'B'$ meet. Likewise, AC and $A'C'$ intersect at a point R, and BC and $B'C'$ intersect at a point P. Since P, Q, R are on extensions of the sides of ABC and $A'B'C'$, they lie in the same plane as each of those two triangles, and consequently lie on the line of intersection of these two planes. Thus P, Q, R lie on a line, which proves the theorem.

The artists who developed the principles of perspective realized that in order to create the appropriate impression of depth in a picture, the painter might have to establish a number of *points at infinity*, where lines in the picture that correspond to parallels in the scene being painted would meet, if extended. They also realized that those points at infinity

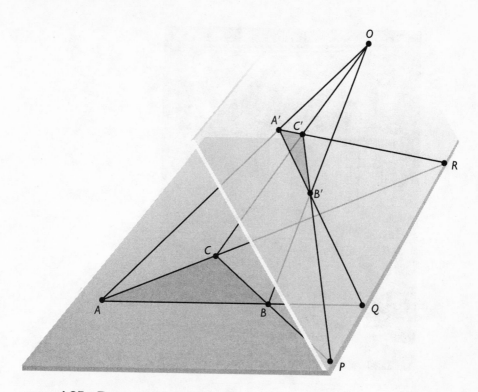

FIGURE **4.25** Desargues' theorem in space.

must all lie on a single line, called the *line at infinity*. This concept is illustrated in Figure 4.26.

Similarly, the mathematicians who developed projective geometry realized that it was convenient to introduce points at infinity. Each line in a plane is assumed to have a single 'ideal' point, a 'point at infinity'. Any two parallel lines are assumed to intersect at their common point at infinity. All of the points at infinity are assumed to lie on a single straight line, the 'ideal line' or 'line at infinity'. This line contains no points other than all the points at infinity.

Notice that only one ideal point is added to each line, not two, as might be suggested by the idea of a line extending to infinity in both directions. In Euclidean geometry, a straight line extends toward (but never reaches) two 'points at infinity' in two directions, but this is not the case in projective geometry: for each line, there is just one point at infinity, and it is on the line.

Ideal points and the ideal line were conceived to avoid issues of parallelism—which is not itself a notion of projective geometry, since projections can destroy parallelism. Because of its Euclidean propensity, the

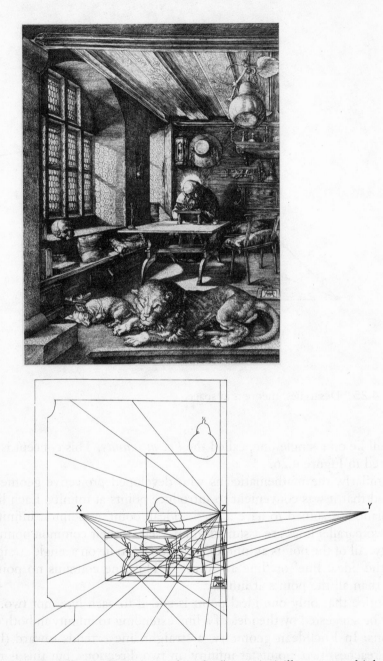

FIGURE **4.26** Albrecht Dürer's woodcut *St. Jerome* (*top*) illustrates multiple-point perspective. An analysis of the perspective (*bottom*) shows that the three 'points at infinity' (*X*, *Y*, and *Z*) all lie on a line—the 'line at infinity'.

human mind cannot readily visualize parallel lines meeting, so it is not possible to completely visualize the process of adding these ideal points and line. The development of projective geometry has to be axiomatic. With the inclusion of ideal points and an ideal line, projective geometry has the following simple axiomatization:

1. There exist at least one point and one line.
2. If X and Y are distinct points, exactly one line passes through them.
3. If l and m are distinct lines, exactly one point is common to them.
4. There are at least three points on any line.
5. Not all points are on the same line.

This axiomatization does not say what a point or a line is, or what it means for a point to lie on a line or for a line to pass through a point. The axioms capture the essential patterns for projective geometry, but do not specify the entities that exhibit those patterns. This is the essence of abstract mathematics.

In this case, one enormous benefit of this level of abstraction is that it practically doubles the number of theorems that can be obtained. Whenever you prove one theorem in projective geometry, a second, so-called *dual theorem* follows automatically, by what is known as the *duality principle*. This principle says that if you take any theorem and everywhere exchange the words 'point' and 'line', exchange the phrases 'points lying on the same line' with 'lines meeting at a single point', and so on, then the resulting statement will also be a theorem, the *dual* to the first theorem.

The duality principle follows from a symmetry in the axioms. First of all, notice that axiom 1 is perfectly symmetrical between points and lines, and that axioms 2 and 3 form a symmetrical pair. Axioms 4 and 5 are not in themselves symmetrical, but if they are replaced with the statements

4'. There are at least three lines through any point.
5'. Not all lines go through the same point.

then the resulting axiom system is equivalent to the first. Thus, any theorem proved from the axioms will still be true if the notions of point and line are interchanged, together with all associated notions.

For example, in the seventeenth century, Blaise Pascal proved the following theorem:

If the vertices of a hexagon lie alternately on two straight lines, the points where opposite sides meet lie on a straight line.

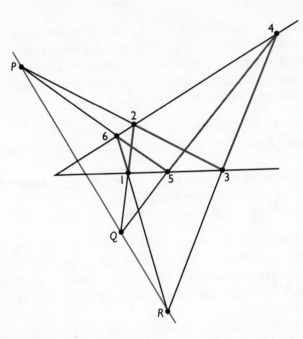

FIGURE **4.27** Pascal's theorem.

This theorem is illustrated in Figure 4.27. A century later, Charles Julien Brianchon proved this theorem:

> If the sides of a hexagon pass alternately through two points, the lines joining opposite vertices meet at a single point.

Brianchon's theorem is illustrated in Figure 4.28. Though Brianchon obtained this result by a separate argument, today's geometer would deduce it at once from Pascal's theorem, as the dual.

The dual of Desargues' theorem is its converse:

> If, in a plane, two triangles are situated so that the corresponding sides, if extended, intersect at three points that lie on a line, then the straight lines joining corresponding vertices will intersect at a single point.

In this case, the duality principle tells us that the converse of Desargues' theorem is true. There is no need for any further proof.

Obviously, the duality principle is possible only if the axioms do not depend on any knowledge of what the words 'point' and 'line' actually mean. When a mathematician is doing projective geometry, he may well have a mental picture consisting of regular points and lines, and he may well draw familiar-looking diagrams to help him solve the problem at hand. But those pictures are to help him carry out the reasoning; the

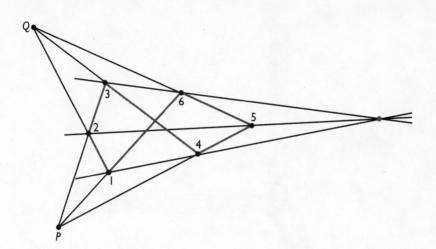

FIGURE **4.28** Brianchon's theorem.

mathematics itself requires no such additional information, and, as David Hilbert suggested, would be equally valid if the words 'point' and 'line' were replaced by 'beer mug' and 'table', and the phrases 'lie on a straight line' and 'meet at a single point' were replaced by 'on the same table' and 'supporting the same beer mug', respectively. This particular change results in a system that does not satisfy the axioms for projective geometry *if these words and phrases are taken to have their usual meanings*. For example, a beer mug can rest on only one table. But if the only properties assumed of these words and phrases are those given by the five axioms for projective geometry, then all the theorems of projective geometry will be true of 'beer mugs' and 'tables'.

Beyond the third dimension

Different geometries represent different ways of capturing and studying patterns of shape. But there are other aspects of shape, whose study involves tools that, while often closely related to geometry, are generally regarded as not being part of geometry itself. One such aspect is dimension.

The concept of dimension is a fundamental one to human beings. The objects that we encounter each day generally have three dimensions: height, width, and depth. Our eyes work as a coordinated pair in order to achieve a three-dimensional view of our surroundings. The theory of perspective described in the previous section was developed in order to create an illusion of this three-dimensional reality on a two-dimensional surface. Whatever the current theories of physics may tell us about the dimensionality of the universe—and some theories involve a

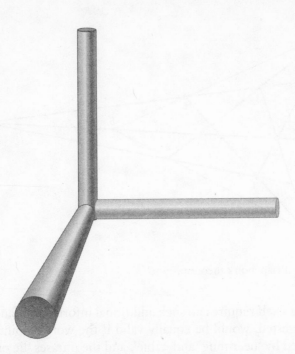

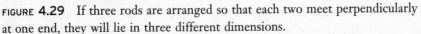

FIGURE **4.29** If three rods are arranged so that each two meet perpendicularly at one end, they will lie in three different dimensions.

dozen or so dimensions—*our* world, the world of our daily experience, is a three-dimensional one.

But what is dimension? What pattern are we capturing when we speak of dimension? One naive description in terms of direction, or straight lines, is illustrated in Figure 4.29. A single straight line represents a single dimension. A second straight line, perpendicular to the first, indicates a second dimension. A third straight line, perpendicular to both of the first two, indicates a third dimension. At this point the process stops, since we cannot find a fourth straight line perpendicular to each of the three we have so far.

A second description of dimension, illustrated in Figure 4.30, is in terms of coordinate systems, as introduced by Descartes. Taken as an axis, a single straight line determines a one-dimensional world. Add a second axis, perpendicular to the first, and the result is a two-dimensional world, or plane. Add a third axis, perpendicular to the first two, and the result is three-dimensional space. Again, the process stops here, since we cannot find a fourth axis perpendicular to the first three.

Both of these approaches are unduly restrictive, in that they are too tightly rooted in the Euclidean notion of a straight line. A better way of capturing the idea of dimension is in terms of degrees of freedom.

A train traveling on a railway track is moving in one dimension. Though the track itself may curve around, climb, and fall, the train has only one direction of motion. (Reversing is regarded as negative movement forward.) The track is embedded in a three-dimensional world, but the train's world is a one-dimensional one. Relative to an origin, the position of the train at any time may be completely specified by just one parameter, the signed distance from the origin measured along the track. One parameter, one dimension.

A ship at sea has two degrees of freedom: forward and backward, left and right. The ship therefore moves in two dimensions. Though the surface of the ocean is roughly spherical, curving around the earth, and thus occupies three dimensions in space, the world the ship moves in is two-dimensional. Its exact position at any time may be specified using just two parameters, its latitude and its longitude. Two parameters, two dimensions.

An airplane in flight moves in a three-dimensional world. The plane can move forward or (after turning around) backward, left or right, up or down. It has three degrees of freedom, and its exact position can be specified using three parameters, latitude, longitude, and altitude. Three parameters, three dimensions.

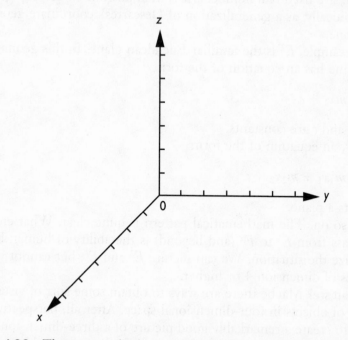

FIGURE 4.30   Three perpendicular axes form the basis for Cartesian geometry in three-dimensional space.

Degrees of freedom do not have to be spatial, as was the case in the examples just considered. Whenever some system varies according to two or more parameters, if it is possible to alter each parameter without changing the others, then each parameter represents a degree of freedom.

When dimensions are regarded not geometrically, but as degrees of freedom, there is no reason to stop at three. For example, the positional status of an airplane in flight can be specified by five parameters: latitude, longitude, altitude, speed, and direction of flight. Each of these can be altered independently of all the others. Suppose we wanted to represent the flight of an airplane in a graphical fashion, as a function of time. The graph would be six-dimensional, associating with each point on the time axis a corresponding 'point' in the five-dimensional space whose axes represent latitude, longitude, altitude, speed, and direction. The result would be a time-dependent 'path' or 'curve' of a 'point' moving in a five-dimensional space.

There is no mathematical restriction on the number of possible axes. For any positive integer n, there is an n-dimensional Euclidean space; it is often denoted by E^n. The n coordinate axes for E^n may be labeled x_1, $x_2, \ldots, x_n$. A point in E^n will be of the form $(a_1, a_2, \ldots, a_n)$, where a_1, $a_2, \ldots, a_n$ are fixed real numbers. It is then possible to develop geometry algebraically, as a generalization of Descartes's coordinate geometry of the plane.

For example, E^2 is the familiar Euclidean plane. In this geometry, a straight line has an equation of the form

$$x_2 = mx_1 + c,$$

where m and c are constants.

In E^3, an equation of the form

$$x_3 = m_1x_1 + m_2x_2 + c$$

represents a plane.

And so on. The mathematical pattern is quite clear. What changes as you pass from E^3 to E^4 (and beyond) is the ability of human beings to visualize the situation. We can picture E^2 and E^3, but cannot visualize spaces of dimension 4 or higher.

Or can we? Maybe there are ways to obtain some sort of visual impression of objects in four-dimensional space. After all, perspective enables us to create a remarkably good picture of a three-dimensional object from a two-dimensional representation. Perhaps we could construct a three-dimensional 'perspective' model of a four-dimensional object, say, a four-dimensional 'hypercube' (the four-dimensional analogue of a cube). Such models have been constructed of wire or metal plates, the better to 'see' the entire structure, and show the three-dimensional 'shadow' cast by the four-dimensional object. Plate 4 presents a photograph of such a model. This particular model presents a 'cube-on view' of a hypercube, the three-dimensional analogue of a two-dimensional, face-on view of a regular cube, shown in the adjacent photograph.

In the case of the regular cube, the face-on view is two-dimensional, so the picture you see is the actual view. The nearest face is a square. Because of the effects of perspective, the face farthest away is a smaller square that sits in the middle of the nearest, and the remaining faces are all distorted into rhombuses.

Just as the faces of a cube are all squares, so the 'faces' of a four-dimensional hypercube are all cubes. A four-dimensional hypercube is bounded by eight 'cube-faces'. Imagine you have the actual, three-dimensional model shown in the photograph in front of you. The large outer cube is the cube-face 'nearest' to you. The smaller cube that sits at the center is the cube-face 'farthest away' from you. The remaining cube-faces are all distorted into rhombic pyramids.

Trying to create a mental image of a four-dimensional object from a three-dimensional model (or a two-dimensional picture of such an object) is by no means an easy matter. Even the reconstruction of a mental image of a three-dimensional object from a two-dimensional picture is a highly complex process in which a number of factors come into play, including context, lighting, shading, shadow, texture, and your expectations.

This is the point made by Plato in his allegory of the cave, in the seventh book of *The Republic*. A race of individuals is constrained from birth to live in a cave. All they know of the outside world consists of the colorless shadows cast on the walls of the cave. Though they are able to gain some sense of the true shapes of solid objects, such as urns, by following the changes in the shadows as the objects are rotated, their mental pictures are destined to remain forever impoverished and uncertain.

Plato's cave was brought up to date in 1978 with the production of the movie *The Hypercube: Projections and Slicings*, by Thomas Banchoff and Charles Strauss at Brown University. The movie was created on a computer, and used color and motion to try to give a sense of the 'true shape' of a four-dimensional hypercube. A frame from this movie is shown in Plate 5.

Because the mental reconstruction of a four-dimensional figure from a three-dimensional model is so difficult, the most reliable way to obtain information about objects such as the hypercube is to forget about trying to picture what is going on and resort to an alternative means of representation: coordinate algebra. Indeed, *The Hypercube* was produced in this way. The computer was programmed to perform the necessary algebra and display the results on the screen in the form of a diagram, using four different colors to represent lines in the four different dimensions.

Using algebra, it is possible to study n-dimensional geometric figures such as n-dimensional hypercubes, n-dimensional hyperspheres, and so forth. It is also possible to use methods of the calculus to study motion in n-dimensional space and to compute the hypervolumes of various n-dimensional objects. For example, using integration, the formula for the volume of a four-dimensional hypersphere of radius r can be computed; it is

$$\frac{1}{2}\pi^2 r^4.$$

The mathematical investigation of figures in four or more dimensions turns out to be more than an intellectual exercise having no applications in the real world. The computer program most widely used in industry is a direct result of such investigations. Called the simplex method, this program tells a manager how to maximize profits—or anything else of interest—in complex situations. Typical industrial processes involve hundreds of parameters: raw materials and components, pricing structures, different markets, a range of personnel, and so forth. Trying to control these parameters so as to maximize profits is a formidable task. The simplex method, invented by the American mathematician George Danzig in 1947, provides a solution based on higher-dimensional geometry. The entire process is represented as a geometric figure in n-dimensional space, where n is the number of independent parameters involved. In a typical case, this figure will look like an n-dimensional version of a polyhedron; such figures are called *polytopes*. The simplex method uses geometric techniques to investigate the polytope so as to find values of the parameters that result in the maximum profit.

Another application of methods that depend on the geometry of higher-dimensional spaces is in routing telephone calls. In this case, the many different ways to route a call from one part of the county to another can be represented in a geometric fashion as a polytope in an n-dimensional space.

Of course, the computers that carry out these calculations cannot 'see' the geometric figures involved. The computer simply performs the algebraic steps it has been programmed to follow. The n-dimensional geometry is used in devising the program in the first place. Mathematicians may be able to express their thoughts using the language of algebra, but generally they do not think that way. Even a highly trained mathematician may find it hard to follow a long algebraic process. But every one of us is able to manipulate mental pictures and shapes with ease. By translating a complicated problem into geometry, the mathematician is able to take advantage of this fundamental human capability.

As I observed at the start of this chapter, there is a sense in which every one of us is a geometer. By combining geometric ideas with the rigorous methods of algebra in the manner just described, the mathematician has been able to make practical use of that observation.

CHAPTER 5

The Mathematics of Beauty

The advantage of working with groups

Geometry sets out to describe some of the visual patterns that we see in the world around us: patterns of shape. But our eyes perceive other patterns: visual patterns not so much of shape per se, but of form. Patterns of symmetry are an obvious example. The symmetry of a snowflake or a flower is clearly related to the obvious geometric regularity of those objects. The study of symmetry captures one of the deeper, more abstract aspects of shape. Because we often perceive these deeper, abstract patterns as beauty, their mathematical study can be described as the mathematics of beauty.

The mathematical study of symmetry is carried out by looking at transformations of objects. To the mathematician, a *transformation* is a special kind of function. Examples of transformations are rotations, translations, reflections, stretchings, or shrinkings of an object. A *symmetry* of some figure is a transformation that leaves the figure invariant, in the sense that, taken as a whole, the figure looks the same after the transformation as it did before, even though individual points of the figure may have been moved by the transformation.

An obvious example of a symmetrical figure is the circle. The transformations that leave the circle invariant are rotations around the center (through any angle, in either direction), reflections in any diameter, or any finite combination of rotations and reflections. Of course, a point marked on the circumference may well end up at a different location: a *marked* circle may possess symmetry neither for rotation nor for reflection. If you rotate a clock face through 90° counterclockwise, the 12 will finish up on the left, where the 9 used to be. So the clock face looks different. Again, if you reflect the clock face in the vertical line that runs from 12 to 6, the 9 and the 3 will swap places, so again, the resulting clock face looks different. Thus, marked circles may possess little symmetry. But the circle itself, ignoring any marks, does have symmetry.

Given any figure, the *symmetry group* of that figure is the collection of all transformations that leave that figure invariant. A transformation in the symmetry group leaves the figure looking exactly the same, in shape, position, and orientation, as it did before.

The symmetry group of the circle consists of all possible combinations of rotations around the center (through any angle, in either direction) and reflections (in any diameter). Invariance of the circle under rotations around the center is referred to as rotational symmetry; invariance with respect to reflections in a diameter is called reflectional symmetry. Both kinds of symmetry are recognizable by sight.

If S and T are any two transformations in the circle's symmetry group, then the result of applying first S and then T is also a member of the symmetry group—since both S and T leave the circle invariant, so does the combined application of both transformations. It is common to denote this double transformation by $T \circ S$. (There is a good reason for the rather perverse-looking order here, having to do with an abstract pattern that connects groups and functions, but I shall not go into that connection here.)

This method of combining two transformations to give a third is reminiscent of addition and multiplication, which combine any pair of integers to give a third. To the mathematician, ever on the lookout for patterns and structure, it is natural to ask what kinds of properties are exhibited by this operation of combining two transformations in the circle's symmetry group to give a third.

First, the combination operation is associative: if S, T, W are transformations in the symmetry group, then

$$(S \circ T) \circ W = S \circ (T \circ W).$$

In this respect, this new operation is very much like the addition and multiplication of integers.

Second, the combination operation has an identity element that leaves unchanged any transformation it is combined with: the null rotation, the rotation through angle 0. The null rotation, call it I, can be applied along with any other transformation T to yield

$$T \circ I = I \circ T = T.$$

The rotation I obviously plays the same role here as the integer 0 in addition and the integer 1 in multiplication.

Third, every transformation has an inverse: if T is any transformation, there is another transformation S such that

$$T \circ S = S \circ T = I.$$

The inverse of a rotation is a rotation through the same angle in the opposite direction. The inverse of a reflection is that very same reflection. To obtain the inverse for any finite combination of rotations and reflections, you take the combination of backward rotations and re-reflections that exactly undoes its effect: start with the last one, undo it, then undo the previous one, then its predecessor, and so on.

The existence of inverses is a property shared with addition of integers: for every integer m there is an integer n such that

$$m + n = n + m = 0$$

(where 0 is the identity for addition); namely, $n = -m$. The same is not true for multiplication of integers, of course: it is not the case that for every integer m, there is an integer n such that

$$m \times n = n \times m = 1$$

(where 1 is the identity for multiplication). In fact, only for the integers $m = 1$ and $m = -1$ is there another integer n that satisfies the above equation.

To summarize, any two symmetry transformations of a circle can be combined by the combination operation to give a third symmetry transformation, and this operation has the three 'arithmetic' properties of associativity, identity, and inverses.

A similar analysis can be carried out for other symmetrical figures. In fact, the properties of symmetry transformations we have just observed in the case of the circle turn out to be sufficiently common in mathematics to be given a name: a *group*. Indeed, I have already used that name in referring to the 'symmetry group'. In general, whenever mathematicians observe some set, G, of entities and an operation, $*$, that combines any two elements x and y in G to give a further element $x * y$ in G, they call this collection a group if the following three conditions are met:

G1. For all x, y, z in G, $(x * y) * z = x * (y * z)$.

G2. There is an element e in G such that $x * e = e * x = x$ for all x in G.

G3. For each element x in G, there is an element y in G such that $x * y = y * x = e$, where e is as in condition G2.

Thus, the collection of all symmetry transformations of a circle is a group. In fact, you should have no difficulty in convincing yourself that if G is the collection of all symmetry transformations of any figure, and * is the operation of combining two symmetry transformations, then the result is a group.

From the remarks made earlier, it should also be clear that if G is the set of integers and the operation * is addition, then the resulting structure is a group. The same is not true for the integers and multiplication. But if G is the set of all rational numbers apart from zero, and * is multiplication, then the result is a group.

A different example of a group is provided by the finite arithmetics discussed in Chapter 1. The integers 0, 1, ..., n − 1 with the operation of addition in modulus n is a group for any integer n. And if n is a prime number, then the integers 1, 2, ..., n − 1 constitute a group under the operation of multiplication in modulus n.

In fact, the three kinds of examples just described barely scratch the surface of the group concept. It turns out to be ubiquitous in modern mathematics, both pure and applied. Indeed, the notion of a group was first formulated, in the early nineteenth century, not in connection with arithmetic or with symmetry transformations, but as part of an investigation of polynomial equations in algebra. The key ideas may be found in the work of Evariste Galois, described later in this chapter.

The symmetry group of a figure is a mathematical structure that in some sense captures the degree of visual symmetry of that figure. In the case of a circle, the symmetry group is infinite, since there are infinitely many possible angles through which a circle may be rotated and infinitely many possible diameters in which it may be reflected. The richness of the circle's group of symmetry transformations corresponds to the high degree of visual symmetry—the 'perfect symmetry'—that we observe when we look at a circle.

At the other end of the spectrum, a figure that is completely unsymmetrical will have a symmetry group that consists of only a single transformation, the identity (or 'do nothing') transformation. It is easy to check that this special case does satisfy the requirements of a group, as does the single integer 0 with the operation of addition.

Before looking at a further example of a group, it is worth spending a few moments reflecting on the three conditions G1, G2, and G3 that determine whether a given collection of entities and an operation constitute a group or not.

The first condition, G1, the associativity condition, is already very familiar to us in the case of the arithmetic operations of addition and multiplication (though not subtraction or division).

Condition G2 asserts the existence of an identity element. Such an element must be unique, for if e and i both have the property expressed by G2, then, applying this property twice in succession, you would have

$$e = e * i = i,$$

so that e and i are in fact one and the same.

This last observation means that there is only one element e that can figure in condition G3. Moreover, for any given element x in G, there is only one element y in G that satisfies the requirement imposed by G3. This is also quite easy to demonstrate. Suppose y and z are both related to x as in G3. That is, suppose that:

(1) $x * y = y * x = e$
(2) $x * z = z * x = e$

Then:

$y = y * e$ (by the property of e)
$\quad = y * (x * z)$ (by equation (2))
$\quad = (y * x) * z$ (by G1)
$\quad = e * z$ (by equation (1))
$\quad = z$ (by the property of e)

so in fact y and z are one and the same. Since there is precisely one y in G related to a given x as in G3, that y may be given a name: it is called the (group) inverse of x, and is often denoted by x^{-1}. And with that, I have just proved a theorem in the mathematical subject known as *group theory*: the theorem that says that, in any group, every element has a unique inverse. I have proved that uniqueness by deducing it logically from the group axioms, the three initial conditions G1, G2, G3.

Though this particular theorem is an extremely simple one, both to state and to prove, it does illustrate the enormous power of abstraction in mathematics. There are many, many examples of groups in mathematics; in writing down the group axioms, mathematicians are capturing a highly abstract pattern that arises in many instances. Having proved, *using only the group axioms*, that group inverses are unique, we know that this fact will apply to every single example of a group. No further work is required. If tomorrow you come across a quite new kind of mathematical structure, and you determine that what you have is a group, you will know at once that every element of your group has a single inverse. In fact, you will know that your newly discovered structure possesses *every* property that can be established—in abstract form—on the basis of the group axioms alone.

The more examples there are of a given abstract structure, such as a group, the more widespread the applications of any theorems proved about that abstract structure. The cost of this greatly increased efficiency is that one has to learn to work with highly abstract structures—with abstract patterns of abstract entities. In group theory, it doesn't matter, for the most part, what the elements of a group are, or what the group operation is. Their nature plays no role. The elements could be numbers, transformations, or other kinds of entities, and the operation could be addition, multiplication, combination of transformations, or whatever. All that matters is that the objects together with the operation satisfy the group axioms G1, G2, and G3.

One final remark concerning the group axioms is in order. In both G2 and G3, the combinations were written two ways. Anyone familiar with the commutative laws of arithmetic might well ask why the axioms were written this way. Why don't mathematicians simply write them one way, say,

$$x * e = x$$

in G2 and

$$x * y = e$$

in G3, and add one further axiom, the commutative law:

G4. For all x, y in G, $x * y = y * x$.

The answer is that this additional requirement would exclude many of the examples of groups that mathematicians wish to consider.

Though many symmetry groups do not satisfy the commutativity condition G4, a great many other kinds of groups do. Consequently, groups that satisfy the additional condition G4 are given a special name: they are called abelian groups, after the Norwegian mathematician Niels Henrik Abel. The study of abelian groups constitutes an important subfield of group theory.

For a further example of a symmetry group, consider the equilateral triangle shown in Figure 5.1. This triangle has precisely six symmetries. There is the identity transformation, I, counterclockwise rotations v and w through 120° and 240°, and reflections x, y, z in the lines X, Y, Z, respectively. (The lines X, Y, Z stay fixed as the triangle moves.) There is no need to list any clockwise rotations, since a clockwise rotation of 120° is equivalent to a counterclockwise rotation of 240° and a clockwise rotation of 240° has the same effect as a counterclockwise rotation of 120°.

There is also no need to include any combinations of these six transformations, since the result of any such combination is equivalent to one

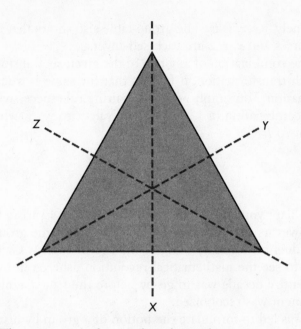

FIGURE **5.1** The symmetries of an equilateral triangle.

of the six given. The table shown in Figure 5.2 gives the basic transformation that results from applying any two of the basic transformations. For example, to find the value of the combination $x \circ v$ from the table, look along the row labeled x and locate the entry in the column labeled v, namely, y. Thus,

$$x \circ v = y$$

in this group. Similarly, the result of applying first w and then x, namely, the group element $x \circ w$, is z, and the result of applying v twice in suc-

The Triangle Symmetry Group

$\circ$	I	v	w	x	y	z
I	I	v	w	x	y	z
v	v	w	I	z	x	y
w	w	I	v	y	z	x
x	x	y	z	I	v	w
y	y	z	x	w	I	v
z	z	x	y	v	w	I

FIGURE **5.2** The triangle symmetry group.

cession, namely, $v \circ v$, is w. The group table also shows that v and w are mutual inverses and x, y, z are each self-inverse.

Since the combination of any two of the given six transformations is another such transformation, it follows that the same is true for any finite combination. You simply apply the pairing rule successively. For example, the combination $(w \circ x) \circ y$ is equivalent to $y \circ y$, which in turn is equivalent to I.

Galois

It is to a brilliant young Frenchman by the name of Evariste Galois that the world owes its gratitude for the introduction of the group concept. Killed in a duel on 30 May, 1832, at the age of twenty, Galois himself never lived to see the mathematical revolution ushered in by his work. In fact, an entire decade was to go by before the true magnitude of his accomplishment was recognized.

Galois was led to formulate the notion of a group by his attempt to solve a specific problem: that of finding simple algebraic formulas for the solution of polynomial equations. Every high-school student is familiar with the formula for the solution of a quadratic equation. The roots of the quadratic equation

$$ax^2 + bx + c = 0$$

are given by the formula

$$x = -b \pm \frac{\sqrt{b^2 - 4ac}}{2a}.$$

Analogous, though more complicated, formulas exist for solving cubic and quartic equations. A cubic equation is one of the form

$$ax^3 + bx^2 + cx + d = 0,$$

and a quartic equation is one that has an additional term involving x^4. The formulas for these equations are analogous in that they involve nothing more complicated than the evaluation of nth roots, or *radicals*. Solutions using such formulas are referred to as solutions by radicals.

In 1824, Abel proved that there can be no such formula for a polynomial of the fifth power, known as a quintic polynomial. More precisely, Abel showed that there is no formula that will work for *all* quintic equations. Among polynomials of degree 5 or more, some can be solved by radicals, but others cannot.

Galois sought a method to determine, for a given polynomial equation, whether or not that equation was solvable by radicals. The task was as ambitious as his solution was original.

Galois discovered that whether an equation can be solved by radicals depends on the symmetries of the equation, and, in particular, on the group of those symmetries. Unless you are a budding Galois, or have seen this before, it probably has not occurred to you that equations can have symmetries, or even that they have any kind of 'shape'. Yet they do. The symmetries of an equation are, it is true, highly abstract, but they are symmetries for all that—not visual symmetries, but algebraic symmetries. Galois took the familiar notion of a symmetry, formulated an abstract means for describing it (namely, symmetry groups), and then applied that abstract notion of symmetry to algebraic equations. It was as brilliant an example of 'technology transfer' as there has ever been.

To get some idea of Galois' reasoning, take the equation

$$x^4 - 5x^2 + 6 = 0.$$

This equation has four roots, $\sqrt{2}$, $-\sqrt{2}$, $\sqrt{3}$, and $-\sqrt{3}$; substituting any one of them for x will produce the answer 0.

In order to forget the numbers themselves and concentrate on the algebraic patterns, call these roots a, b, c, and d, respectively. Clearly, a and b form a matching pair, as do c and d. In fact, this similarity goes much deeper than b being equal to $-a$ and d being equal to $-c$: there is an 'algebraic symmetry' between a and b and between c and d. Any polynomial equation (with rational coefficients) satisfied by one or more of a, b, c, and d will also be satisfied if we swap a and b or if we swap c and d, or if we make both swaps at once. If the equation has just one unknown x, swapping, say, a and b will simply amount to replacing a with b in the equation, and vice versa. For instance, a satisfies the equation $x^2 - 2 = 0$, and so does b. In the case of the equation $x + y = 0$, the values $x = a, y = b$ is a solution, and so is $x = b, y = a$. For an equation with four unknowns, w, x, y, z, that is solved by a, b, c, d, it is possible to make two genuine swaps at once. Thus, a and b are indistinguishable, and so are c and d. On the other hand, it is easy to distinguish, say, a from c. For example, $a^2 - 2 = 0$ is true, but $c^2 - 2 = 0$ is false.

The possible permutations of the four roots—swapping a and b, swapping c and d, or performing both swaps together—constitute a group known as the Galois group of the original equation. It is the group of symmetries with respect to polynomial equations (with rational coefficients and one or more unknowns) satisfied by the four roots. For obvious reasons, groups that consist of permutations, of which Galois groups are one kind of example, are referred to as permutation groups.

Galois found a structural condition on groups—that is to say, a property that some groups will possess and others will not—such that the original equation will have a solution by radicals if, and only if, the Galois group satisfies that structural condition. Moreover, Galois' structural condition depends only on the arithmetical properties of the group. Thus, in principle, whether or not a given equation can be solved by radicals can be determined solely by examining the group table of the Galois group.

To put the historical record straight, Galois did not formulate the concept of an abstract group in the clean, crisp way it was presented above, in terms of the three simple axioms G1, G2, G3. That formulation was the result of efforts by Arthur Cayley and Edward Huntington around the turn of the century. But the essential idea was undoubtedly to be found in Galois' work.

Once Galois' ideas became known, a number of mathematicians developed them further and found other applications. Augustine-Louis Cauchy and Joseph Louis Lagrange carried out investigations of permutation groups, and in 1849, Auguste Bravais used groups of symmetries in three-dimensional space in order to classify the structures of crystals, establishing a close interplay between group theory and crystallography that continues to this day (see page 215).

Another major stimulus to the development of group theory was provided in 1872, when Felix Klein, lecturing at Erlangen, in Germany, established what became known as the 'Erlangen program', an attempt, almost entirely successful, to unify geometry as a single discipline. The discussions in the previous chapter should indicate why mathematicians felt the need for such a unification. In the nineteenth century, after the two-thousand-year reign of Euclidean geometry, there had suddenly appeared a whole range of different geometries: Euclidean, Bolyai-Lobachevsky, Riemannian, projective, and several others, including the most recent arrival—and the most difficult to swallow as a 'geometry'—topology, discussed in the next chapter.

Klein proposed that *a geometry* consists of the study of those properties of figures that remain invariant under a certain group of transformations (of the plane, of space, or whatever). For example, Euclidean geometry of the plane is the study of those properties of figures that remain invariant under rotations, translations, reflections, and similarities. Thus, two triangles are congruent if one may be transformed into the other by means of a 'Euclidean symmetry', a combination of a translation, a rotation, and possibly also a reflection. (Euclid's definition was that two triangles are congruent if they have corresponding sides equal in length.) Similarly, projective geometry of the plane is the study of those properties of figures that remain invariant under members of the group of projective transformations of the plane. And topology is the study of those properties of figures left unchanged by topological transformations.

With the success of the Erlangen program, a further level of abstract pattern was uncovered: the pattern of different geometries. This highly abstract pattern was described by means of the group-theoretic structures of the groups that determine the geometries.

How to stack oranges

Mathematical patterns can be found everywhere. You see symmetry whenever you gaze at a snowflake or a flower. A trip to your local supermarket can provide another kind of pattern. Take a look at the piles of oranges (see Figure 5.3). How are they arranged? How were they arranged in the crates in which they were shipped? In the case of the fruit display, the aim is to create an arrangement that is stable; for shipping, the desire is to pack the maximum number of oranges into a given container. Aside from the rather obvious fact that the absence of sidewalls means that the display pile must have some sort of overall pyramid form, are the two arrangements the same? Do stability and efficiency of packing lead to the same arrangement? If they do, can you explain why?

FIGURE 5.3 The familiar pattern of a stack of oranges in a supermarket.

In other words, is the familiar arrangement of oranges in a super-market the most efficient? That is, does it maximize the number of or-anges that can be fitted into the available space?

Like symmetry, the patterns involved in packing objects can be stud-ied mathematically. The mathematician's version of the supermarket manager's problem of packing oranges efficiently is known as *sphere packing*. What is the most efficient way to pack identical spheres? De-spite a history of investigations into the subject that goes back at least as far as work by Kepler in the seventeenth century, some of the most basic questions about sphere packing remain unanswered to this day.

Thus forewarned, it is perhaps wise to pull back momentarily from the thorny question of sphere packing to what must surely be an easier problem: the two-dimensional analogue of circle packing. What is the most efficient way to pack identical circles (or disks) in a given area?

The shape and size of the area to be filled can surely make a differ-ence. So, in order to turn this question into one that is precise and math-ematical, you should do what mathematicians always do in such cir-cumstances: fix on one particular case, chosen so as to get to what seems to be the heart of the problem. Since the point at issue is the pattern of the packing, not the shape or size of the container, you should concen-trate on the problem of filling the whole of space—two-dimensional space in the case of disks, three-dimensional space in the case of spheres. Any answers you obtain in this mathematician's idealization will pre-sumably hold approximately for sufficiently large real-life containers. The larger the container, the better the approximation will be.

Figure 5.4 indicates the two most obvious ways of packing disks, called the rectangular and the hexagonal arrangements. (This terminol-ogy refers to the figures formed by tangents common to the adjacent cir-cles.) How efficient are these two arrangements at packing the maximum number of disks? The decision to concentrate on packing the entire plane means that you have to be careful how you formulate this ques-tion in precise terms. The quantity that measures the efficiency of any packing is surely the *density*, the total area or volume of the objects be-ing packed divided by the total area or volume of the container. But when the container is all of the plane or space, calculation of the den-sity in this fashion will produce the nonsensical answer ∞/∞.

The way out of this dilemma can be found in the methods described in Chapter 3. The density of packing can be calculated using the same kind of pattern that gave Newton and Leibniz the key to the calculus: the method of limits. You define the density of a packing arrangement by first calculating the ratio of the total area or volume of the objects packed to the area or volume of the container, for larger and larger

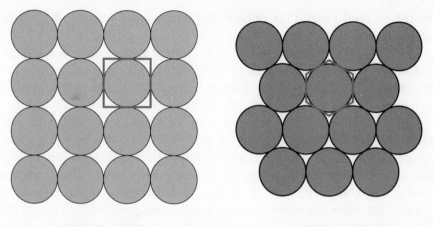

Rectangular packing　　　　　　　　　Hexagonal packing

FIGURE **5.4** Two ways of packing identical disks to fill the plane: the 'rectangular' arrangement and the 'hexagonal' arrangement.

finite containers. You then compute the limit of those ratios as the containers' boundaries tend toward infinity (i.e., get larger without bound). Thus, for example, in the case of disk packing, you can compute the (finite) density ratios for packings that attempt to fill larger and larger square regions whose edges increase without bounds. Of course, as in the case of the calculus, there is no need to compute an endless series of actual, finite ratios; rather, you look for the appropriate pattern in the form of a formula and then compute the limit by looking at that pattern.

In the case of disk packing in the plane, this strategy was followed by Kepler, who found that the density of the rectangular arrangement is $\pi/4$ (approximately 0.785) and that of the hexagonal arrangement $\pi/2\sqrt{3}$ (approximately 0.907). The hexagonal arrangement gives the greater density.

Of course, this last conclusion is hardly a surprise: a brief glance at Figure 5.4 indicates that the hexagonal arrangement leaves less space between adjacent disks, and is therefore the more efficient of the two. But it is not quite as obvious that the hexagonal packing is *the* most efficient—that is to say, has a greater density than *any* other packing. The problem with this more general question is that it asks about all possible arrangements of disks, however complex, be they regular or irregular. In fact, so difficult does it seem to resolve the question of what is the most efficient disk packing of all, that it is sensible to look first at a further special case, one that introduces far more pattern than is present in the general case. This is how mathematicians actually did finally arrive at a solution.

In 1831, Gauss showed that the hexagonal arrangement is the densest among the so-called lattice packings. It was Gauss's concept of a *lattice* that provided the crucial additional structure needed, enabling some progress to be made.

In the plane, a lattice is a collection of points arranged at the vertices of a regular, two-dimensional grid. The grid may be square, rectangular, or in the form of identical parallelograms, as illustrated in Figure 5.5. In mathematical terms, the crucial feature of a (planar) lattice is that it has translational invariance, or translational symmetry. This means that certain translations of the plane, without rotation, leave the entire lattice superimposed upon its original position, so that it appears to be unchanged.

A lattice packing of disks is one in which the centers of the disks form a lattice. Such packings are obviously highly ordered. The rectangular and hexagonal disk packings are clearly lattice packings. By relat-

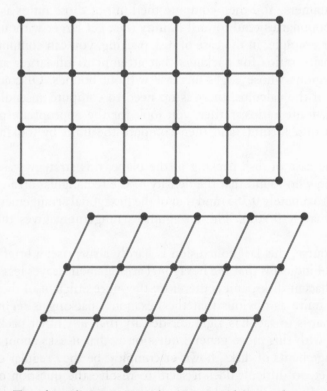

FIGURE **5.5** Two planar lattices, one with a square grid (*top*) and one in the form of identical parallelograms (*bottom*).

ing lattice packings of disks in the plane to number theory and making use of some number-theoretic work that had been done by Lagrange, Gauss determined that the hexagonal is the most efficient lattice packing.

Of course, Gauss's discovery left open the question of whether or not the hexagonal arrangement is the most efficient of all disk packings, regular or otherwise. In 1892, Axel Thue announced that he could show that the answer was yes, though it was not until 1910 that he presented a reasonably complete proof.

With the two-dimensional case disposed of, what about the original, three-dimensional sphere-packing problem? Again, it makes sense to do what Gauss did and concentrate first of all on lattice packings, in which the centers of the spheres form a three-dimensional lattice—a regular, three-dimensional grid.

There are, it turns out, exactly fourteen different kinds of three-dimensional lattices. This result was finally established by the French botanist and physicist Auguste Bravais in 1848, building upon the work of a number of mathematicians. Sometimes referred to as Bravais lattices, they are shown in Plate 6.

One obvious way of arranging spheres in a regular, lattice fashion is to build up the arrangement layer by layer, much as the assistant in the supermarket stacks the oranges. In order to obtain an efficient packing, it seems reasonable to arrange each layer so that the centers of the spheres are in one of the two planar lattice formations considered above, the rectangular or the hexagonal. Depending on how you do this, you can obtain three different three-dimensional arrangements, shown in Figure 5.6. They arise as follows.

If a rectangular formation is chosen for the layers, there are two ways of stacking the layers: with the spheres in each layer directly above one another, or staggered so that each sphere in the upper level nestles between four spheres beneath it. (This second alternative is the one used—for stability—in order to stack oranges.) In the former arrangement, the centers of the spheres form a cubic lattice; in the latter, the centers constitute what is known as a face-centered cubic lattice, a cubic lattice in which each cube is 'stood on one corner'.

Alternatively, the layers can consist of hexagonal lattice formations, and again, there are two ways in which the layers may be stacked one upon another, aligned and staggered. This would appear to give a total of four different three-dimensional lattice packings, but in fact there are only three. Staggered layers of hexagonally packed spheres and staggered layers of squarely packed spheres are equivalent, in that one is just the other viewed from a different angle. You can see this equivalence for

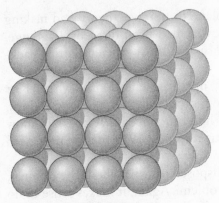

Cubic lattice

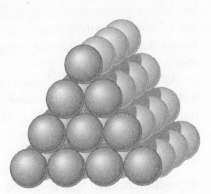

Face-centered cubic lattice

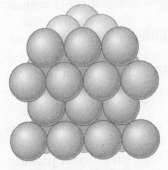

Hexagonal lattice

FIGURE 5.6 Three different ways to arrange spheres by stacking regular layers. The centers of the spheres constitute a cubic, a face-centered cubic, and a hexagonal lattice, respectively.

yourself quite easily by looking at a pile of oranges arranged in the familiar staggered square pyramid fashion. If you look at one of the slanting faces, you will see one layer of what is a staggered hexagonal packing.

In the third distinct lattice packing, the aligned hexagonal packing, the centers of the spheres constitute what is known as a three-dimensional hexagonal lattice.

Kepler computed the density associated with each of these three lattice packings and obtained the figures $\pi/6$ (approximately 0.5236) for the cubic lattice, $\pi/3\sqrt{3}$ (approximately 0.6046) for the hexagonal lattice, and $\pi/3\sqrt{2}$ (approximately 0.7404) for the face-centered cubic lattice. Thus, the face-centered cubic lattice—the orange-pile arrangement—is easily the most efficient packing of the three. But is it the most efficient of *all* lattice packings? More generally, is it the most efficient of all packings, regular or otherwise?

The first of these two questions was answered by Gauss, not long after he solved the analogous problem in two dimensions. Again, he reached his answer by using results from number theory. But the second problem remains unsolved to this day. We simply do not know for sure whether the familiar orange-pile arrangement of spheres is the most efficient of all arrangements.

The orange pile is certainly not the *unique* best packing, since there are non-lattice packings with exactly the same density. It is easy to construct such a non-lattice packing by stacking hexagonal layers as follows: Put down one hexagonal layer. Lay a second on the first in a nestled fashion. Now add a third layer, nestled on the second. There are two distinct ways to do this. In one way, the spheres in the third layer lie directly above corresponding spheres in the first; in the other way, there is no such alignment. The second of these two alternatives will, if repeated, lead to the face-centered cubic lattice. The first alternative leads to a non-lattice packing that has the same density as the second.

Indeed, by stacking hexagonal layers and randomly choosing between the two alternatives at each stage, you can obtain a sphere packing that is random in the vertical direction, and yet has exactly the same density as the face-centered cubic arrangement.

Before leaving the topic of sphere packing, I should mention that, although mathematicians do not know for sure that supermarkets pile oranges in the most efficient fashion, they do know that the arrangement used is very close to the best. It has been proved that no sphere packing can have a density greater than 0.77836.

Snowflakes and honeycombs

Kepler's interest in sphere packing, it has to be said, was not driven by an overpowering interest in piles of fruit, but it was, nonetheless, motivated by an equally real phenomenon, the shape of a snowflake. Moreover, Kepler's work did involve fruit: not the mundane orange, but the (mathematically) far more interesting pomegranate. Honeycombs figured in his investigations as well.

As Kepler observed, although any two snowflakes may differ in many minute ways, they all share the mathematical property of sixfold, or hexagonal, symmetry: if you rotate a snowflake through 60° (one-sixth of a complete rotation), then, like a hexagon, it will appear unchanged (see Figure 5.7). Why is it, Kepler asked, that all snowflakes have this fundamental hexagonal form?

As usual, Kepler sought his answer in geometry. (Remember, his most famous result was the discovery that the planets follow elliptical

FIGURE **5.7** The sixfold symmetry of a snowflake. If you rotate a snowflake by any multiple of 60° (one-sixth of a complete rotation), it always will look the same.

orbits. Remember too his fascination with Plato's atomic theory, which was based on the five regular solids, and his attempt to describe the solar system in terms of those same geometric figures, described on page 150.) His idea was that natural forces impose a regular geometric structure on the growth of such seemingly diverse objects as snowflakes, pomegranates, and honeycombs.

Kepler's key structural notion was of geometric solids that could be fitted together so as to completely fill space. A natural way to obtain such figures, he suggested, was to start with an arrangement of spheres and imagine that each sphere expands so as to completely fill the intermediate space. Assuming that nature always adopts the most efficient means to achieve her ends, the regular patterns of the honeycomb and the pomegranate and the hexagonal shape of the snowflake could all be explained by examining efficient packings of spheres and observing the geometric solids they give rise to.

In particular, spheres arranged in a cubic lattice will expand into cubes, spheres arranged in a hexagonal lattice will expand to give hexagonal prisms, and spheres arranged in a face-centered cubic lattice will expand to give Kepler's so-called rhombic dodecahedra (see Figure 5.8). Indeed, in the case of the pomegranate, these theoretical conclusions seem to be substantiated in fact: the seeds in a growing pomegranate are initially spherical, arranged in a face-centered cubic lattice. As the pome-

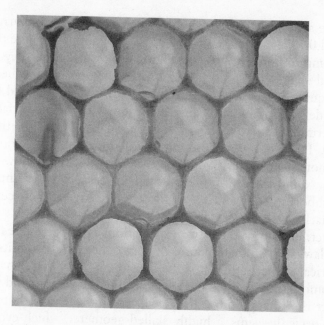

FIGURE **5.8** A rhombic dodecahedron, a figure having twelve identical rhombic faces.

granate grows, the seeds expand until they achieve the form of rhombic dodecahedra that completely fill the internal space.

In addition to stimulating mathematical investigations into sphere packing, Kepler's ideas led to a number of experimental studies of packing. For instance, the delightfully titled *Vegetable Staticks*, a 1727 work by an Englishman named Stephen Hales, describes how he filled a pot with peas and compressed them as far as possible, whereupon he observed that the peas had each assumed the shape of a regular dodecahedron. In fact, Hales's claim was presumably exaggerated, since regular dodecahedra do not pack space. However, though a random initial arrangement of spheres will not lead to regular dodecahedra, it will produce various kinds of rhombic shapes.

In 1939, two botanists, J. W. Marvin and E. B. Matzke, arranged lead shot in a steel cylinder in the familiar orange-stacking manner and then compressed the contents with a piston, obtaining the theoretically predictable rhombic dodecahedra of Kepler. Repeating the experiment with randomly packed shot led to irregular, fourteen-sided figures.

Further experimental work of this nature also demonstrated that random packings are, in general, not as efficient as the face-centered cubic lattice arrangement. The highest density for a random arrangement is around 0.637, as opposed to 0.740 for the familiar orange pile.

Turning to honeycombs (see Figure 5.9), to what do they owe their hexagonal shape? It would be reasonable to suppose that bees secrete wax in liquid form, which then forms itself into the observed hexagonal shape under the influence of surface tension. Minimizing surface tension would certainly lead to a hexagonal lattice shape. This is the theory proposed by D'Arcy Thompson, author of the celebrated book *Laws of Form*. Another possibility would be that the bees first hollow out cylinders of wax and then push out the walls of each cylinder until they hit the neighboring cells and fill in the empty space in between. This is what Charles Darwin thought happened.

In fact, neither of these two explanations is correct. The fact is, it is not the laws of inanimate nature that give the honeycomb its elegantly symmetrical shape; rather, it is the bees themselves that construct their honeycomb in this fashion. The bees secrete the wax as solid flakes, and construct the honeycomb cell by cell, face by face. The humble bee is in some ways, it seems, a highly skilled geometer, which evolution has equipped for the task of constructing its honeycomb in the form that is mathematically the most efficient.

Finally, what of the snowflake that motivated Kepler to begin his initial study of sphere packing? During the years after Kepler, scientists gradually came to believe that he had been correct, and that the regular, symmetrical shape of crystals reflected a highly ordered internal structure. In 1915, using the newly developed technique of X-ray dif-

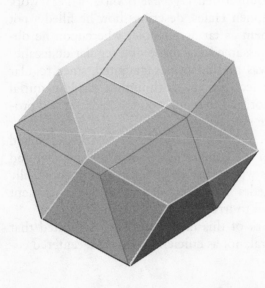

FIGURE **5.9** A close-up view of a honeycomb.

fraction, Lawrence Bragg was able to demonstrate this conclusively. Crystals do indeed consist of identical particles (atoms) arranged in regular lattices (see Figure 5.10).

The snowflake starts out as a tiny, hexagonal crystal seed of ice in the upper atmosphere. As air currents carry it up and down through different altitudes at different temperatures, the crystal grows. The actual pattern that results depends on the particular movements of the growing snowflake through the atmosphere, but since the snowflake is small, the same pattern of growth occurs on all sides, and hence the hexagonal shape of the original seed crystal is preserved. This process gives rise to the familiar hexagonal symmetry. (Incidentally, the word *crystal* comes from the Greek word for 'ice'.)

It should be clear by now that the mathematical study of sphere packing can contribute to our understanding of certain phenomena in the world around us. This was, after all, Kepler's reason for commencing such a study. What he most certainly did not anticipate was that the study of sphere packing would have an application in the twentieth-century technology of digital communications, which would arise as a result of generalizing the sphere-packing problem to spaces of dimension 4 and more! This surprising, and fairly recent, development is yet another example of a practical application arising from the pure mathematician's search for abstract patterns that exist only in the human mind.

In four and five dimensions, the densest lattice packing of space is the analogue of the face-centered cubic lattice packing, but for dimen-

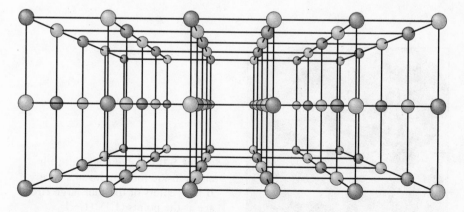

FIGURE **5.10** Crystals of common salt (sodium chloride) form perfect cubes. Their external form reflects the internal structure of the salt molecule, which is a cubic lattice made up of sodium ions (darker spheres) and chloride ions (lighter spheres), alternating in all directions.

sions greater than 5, that is no longer the case. The crucial factor is that, as the number of dimensions increases, there is more and more space between the various hyperspheres. By dimension 8, there is so much free space in the face-centered cubic lattice packing that it is possible to fit a second copy of the same packing into the available gaps without any of the spheres overlapping. The packing that results is the densest lattice packing in eight dimensions. Moreover, certain cross sections of this packing are the densest lattice packings in six and seven dimensions. These results were discovered by H. F. Blichfeldt in 1934.

To get a sense of sphere packings in higher-dimensional spaces, let's look at some simple cases.

Figure 5.11 shows four circles, each of radius 1, packed snugly into a 4×4 square. Adjacent circles just touch each other. Clearly, it is possible to fit in a smaller, fifth circle, centered at the origin, so that it just touches the original four circles.

Figure 5.12 illustrates the analogous situation in three dimensions. Eight spheres of radius 1 can be packed tightly into a $4 \times 4 \times 4$ cubic box. A smaller, ninth sphere can obviously be fitted into the center so that it just touches each of the original eight spheres.

Though you cannot visualize it, the same can be done in four, five, or indeed, any number of dimensions. For instance, you can pack sixteen four-dimensional hyperspheres of radius 1 into a four-dimensional hypercube that measures $4 \times 4 \times 4 \times 4$, and you can fit an additional hypersphere into the center so that it just touches each of the original hyperspheres. There is an obvious pattern: in dimension n, you can pack 2^n hyperspheres of radius 1 into a hypercube whose edges are all of

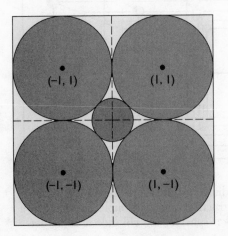

FIGURE **5.11** Packing a small fifth circle inside four circles in a square. If the centers of the four large circles are at the points $(1,1)$, $(1,-1)$, $(-1,-1)$, and $(-1,1)$, then the fifth circle has its center at the origin, as shown.

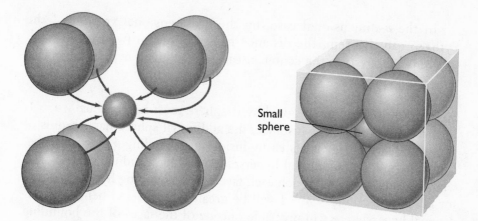

FIGURE **5.12** Packing a small ninth sphere inside eight spheres in a cube.

length 4, and then you can pack an additional hypersphere into the center so that it touches all of them.

Let us see what other patterns can be found. In the original two-dimensional case, the additional fifth circle is contained inside the bounding outer square, and in the three-dimensional case, the additional ninth sphere is contained inside the bounding cubic container.

Similarly, in dimensions 4, 5, and 6, the additional hypersphere sits inside the hypercube that contains the original hyperspheres. This pattern seems so clear that it would be a rare individual who did not assume that the same is true in any number of dimensions. But there is a surprise in store. When you reach dimension 9, something strange happens: the additional nine-dimensional hypersphere actually touches each face of the bounding hypercube; and in dimension 10 and above, the additional hypersphere actually protrudes outside the hypercube. The amount of protrusion increases as the dimension goes up.

This surprising conclusion is easily arrived at using algebra. In the two-dimensional case, what must the radius of the additional circle be in order for it to just touch the original four disks? By the Pythagorean theorem, the distance from the origin to the center of each of the four given circles is

$$\sqrt{1^2 + 1^2} = \sqrt{2}.$$

Since each original circle has radius 1, the additional circle must have radius $\sqrt{2} - 1$, which works out to about 0.41. Obviously, an additional circle of radius 0.41 whose center is at the origin will fit easily into the original 4×4 square container.

In the n-dimensional case, by the n-dimensional version of the Pythagorean theorem, the distance from the origin (where the small sphere is centered) to the center of each of the given hyperspheres is

$$\sqrt{1^2 + 1^2 + \ldots + 1^2} = \sqrt{n}.$$

So the additional n-dimensional hypersphere must have radius $\sqrt{n} - 1$. In three dimensions, for instance, the additional sphere has a radius of about 0.73, which again fits easily inside the bounding $4 \times 4 \times 4$ cube. But for $n = 9$, the additional hypersphere will have radius $\sqrt{9} - 1 = 2$, which means it will just touch each face of the bounding hypercube, and for $n > 9$, the radius $\sqrt{n} - 1$ will be greater than 2, and hence the additional hypersphere will protrude outside of the faces of the bounding hypercube.

The point is, the greater the number of dimensions, the more room there is between the original hyperspheres. Even going from two to three dimensions, the additional 'hypersphere' becomes closer to the bounding 'hypercube': the radius increases from 0.41 to 0.73. The eventual protrusion of the additional hypersphere is surprising only because it does not happen until dimension 9, which is outside of our everyday experience.

The application of sphere packing to communications technology, which I mentioned earlier, arises when you consider sphere packing in twenty-four-dimensional space. In 1965, John Leech constructed a remarkable lattice packing based on what is now known as the Leech lattice. This lattice, which has deep connections to group theory, gives a sphere packing that is almost certainly the densest lattice packing in twenty-four-dimensional space, with each hypersphere touching 196,560 others. The discovery of the Leech lattice led to a breakthrough in the design of what are known as error-detecting codes and error-correcting codes for data transmission.

Though surprising at first glance, the connection between sphere packing and the design of data codes is a simple one. (In order to illustrate the general idea, I shall, however, greatly simplify matters.) Imagine you are faced with designing a means of encoding distinct words in digital form in order to transmit messages over some communications network. Suppose you decide to use eight-bit binary strings for your codes, so that each word is coded by a string such as (1,1,0,0,0,1,0,1), (0,1,0,1,1,1,0,0), and so on. During transmission of the signal, interference on the channel might cause one or two bits to be miscommunicated, so that an originating string (1,1,1,1,1,1,1,1) arrives as, say, (1,1,1,1,0,1,1,1). In order to be able to detect such a communication error, it would be sensible to design your coding scheme so that the second of these two strings was not the code for any word, and hence would be recognized as having resulted from a miscommunication. Even better, if you could then recognize what the originating string must most likely have been in order to result in such an arriving string, you could correct the error. In order to achieve these two aims, you need to choose your coding strings so that any two strings that are used as codes differ from each other in at least three binary places. On the other hand, in order to be able to encode all the messages that need to be transmitted, you need to have the largest possible stock of coding strings. If you now view the problem geometrically, what you have is a sphere-packing problem in eight-dimensional space.

To see this, imagine that all possible code words are points in eight-dimensional space. This collection clearly constitutes a (hyper-)cubic lattice, the lattice of all points whose coordinates are 0 or 1. For each string s chosen as a code word, you want to ensure that no other string chosen as a code word differs from s by three or fewer bits. In geometric terms, what this amounts to is that no lattice point that lies within a sphere of radius $r = \sqrt{3}$, centered at s, is also a coding string. Maximizing the number of coding strings without having two coding strings that come within a distance r of each other is thus equivalent to finding the densest packing of spheres of radius $r/2$ on that lattice.

And there you have it: from snowflakes and pomegranates to modern telecommunications techniques, by way of the geometry of multi-dimensional space!

How many wallpaper patterns are there?

In contrast to digital communications, the design of wallpaper patterns might sound frivolous, but if a study of snowflakes and pomegranates can lead to the design of error-correcting codes, there is no telling where an investigation of wallpaper patterns might lead. Certainly, the mathematics of wallpaper patterns turns out to be deep and of considerable intrinsic interest.

The characteristic of a wallpaper pattern of most interest to the mathematician is that it repeats in a regular fashion to completely fill the plane (see Plate 7). Thus, the 'wallpaper patterns' the mathematician studies may also be found on linoleum floors, patterned cloth, rugs and carpets, and so forth. In these real-life examples, the pattern repeats until the wall, floor, or material ends; the mathematician's pattern stretches out to infinity in every direction.

The mathematical idea behind any wallpaper pattern is that of the *Dirichlet domain*. Given any point in a planar lattice, the Dirichlet domain of that point (for the given lattice) consists of the entire region of the plane that is nearer to that point than to any other lattice point. The Dirichlet domains of a lattice provide a 'brick model' of the symmetry of the lattice.

To design a new wallpaper pattern, all you need to do is produce a pattern that will fill one portion of the paper, and then repeat that pattern over all such portions. More precisely, you start off with a lattice grid, fill in one particular Dirichlet domain with your pattern, and then repeat the same pattern in all other Dirichlet domains. Even if the designer does not consciously follow this approach, it is a fact that any wallpaper pattern can be regarded as having been produced in this manner.

There are just five distinct kinds of Dirichlet domains that can arise in the plane, each one either a quadrilateral or a hexagon. These five types are shown in Figure 5.13.

There is, of course, no limit to the different kinds of wallpaper patterns that can be designed. But how many of the patterns produced in this way are mathematically distinct, in the sense of having different symmetry groups? The answer may surprise you. For all that there is no

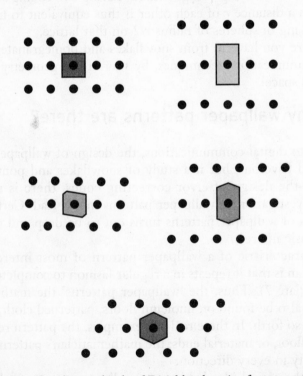

FIGURE **5.13** The five distinct kinds of Dirichlet domains for a two-dimensional lattice.

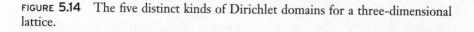

FIGURE **5.14**   The five distinct kinds of Dirichlet domains for a three-dimensional lattice.

limit to the number of different patterns that can be drawn, each pattern will be one of only seventeen distinct kinds in terms of symmetry groups. This is because there are exactly seventeen distinct groups that correspond to symmetries of wallpaper patterns. The proof of this fact is quite difficult. The patterns shown in Plate 8 illustrate the seventeen different symmetry types. Studies of the various repeating patterns that artists and designers have utilized over the course of history have uncovered examples of all seventeen possibilities.

The notions of Dirichlet domains and of wallpaper patterns obviously generalize to three dimensions. The fourteen types of three-dimensional lattices illustrated in Plate 6 give rise to precisely five distinct kinds of Dirichlet domains, shown in Figure 5.14. These five figures constitute a collection of solids every bit as fundamental as the five Platonic solids, though they are far less well known.

The five Dirichlet domains give rise to exactly 230 different symmetry groups for three-dimensional wallpaper patterns. Many of these three-dimensional patterns arise in nature, in the structure of crystals. Consequently, the mathematics of symmetry groups plays an important role in crystallography. Indeed, most of the work on classifying the 230 different symmetry groups was carried out by crystallographers during the late nineteenth century.

How many ways can you tile a floor?

While sphere packing addresses the problem of finding the optimal way of arranging a particular shape—a sphere—to achieve the greatest density, the mathematical study of tilings looks at a slightly different problem: *what* shapes can be stacked together to fill space completely? This fundamental question is analogous to investigating the manner in which matter splits up into atoms and natural numbers split up into products of primes.

For instance, starting with the two-dimensional case, squares, equilateral triangles, and hexagons can each be arranged to completely fill, or *tile*, all of the plane, as illustrated in Figure 5.15. Are these the only regular polygons that can tile the plane? (A *regular* polygon is one for which all sides are the same length and all the interior angles are equal.) The answer turns out to be yes.

If you allow two or more kinds of tiles, but impose the additional requirement that the same array of polygons surround each vertex, then there are exactly eight further possibilities, made up of combinations of triangles, squares, hexagons, octagons, and dodecagons, as illustrated in Plate 9. The regularity of these eleven tilings is sufficiently pleasing to the eye that each will make an attractive pattern for tiling a floor, perhaps to complement one of the seventeen basic wallpaper patterns considered earlier.

If nonregular polygons are allowed, then there is no limit to the number of possible tilings. In particular, any triangle or any quadrilateral can be used to tile the plane. Not every pentagon will work, however; indeed, a tiling consisting of *regular* pentagons will not completely fill the

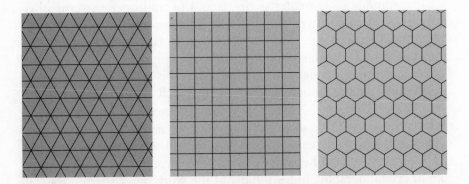

FIGURE 5.15 There are three regular polygons that con be arranged to completely tile the plane: the equilateral triangle, the square, and the regular hexagon.

plane, but will leave gaps. On the other hand, any pentagon having a pair of parallel sides will tile the plane. To date, mathematicians have identified fourteen distinct kinds of pentagons that will tile the plane, the last category being found as recently as 1985, but it is not known whether this list is complete. (This finding needs some qualification: it applies only to convex pentagons, those in which all corners point outward.)

In the case of hexagons, it was proved back in 1918 that there are precisely three kinds of convex hexagons that will tile the plane; these are shown in Plate 10. With hexagons, the possibility of tiling the plane using a single kind of nonregular convex polygon comes to an end. No convex polygon having seven or more sides can be arranged to completely fill the plane.

As in the case of sphere packing, in which initial studies separated lattice packings from irregular ones, it is possible to split tilings into two types. One type comprises tilings that cover the plane in a repeating (or *periodic*) pattern that has translational symmetry, like those illustrated in Figure 5.15. The other type comprises those tilings having no translational symmetry, the *aperiodic* tilings. The distinction between periodic and aperiodic tilings is somewhat analogous to the distinction between rational and irrational real numbers, in which the latter have decimal expansions that continue forever, without settling into repeating blocks.

Quite a lot is known about the periodic tilings. In particular, any such tiling of the plane must have as its symmetry group one of the seventeen groups of wallpaper patterns discussed in the previous section. But what of the aperiodic patterns? Indeed, are there any such tilings? Does the plane split up into pieces that have the same shape other than in a periodic fashion? In 1974, British mathematician Roger Penrose gave the—perhaps somewhat surprising—answer: yes. Penrose discovered a pair of polygons that could be fitted together to completely fill the plane, but only aperiodically—that is to say, without translational symmetry. Penrose's original pair of polygons were not both convex, but two convex polygons that will also tile the plane in an aperiodic fashion were found subsequently, and are shown in Figure 5.16. (The new pair is closely related to the original pair found by Penrose.) As was the case with Penrose's original tiles, in order to force the aperiodicity of any tiling using such figures, the edges of the polygons—both rhombuses—have to be assigned a specific direction, and the tiling has to be constructed so that the directions match up along any join.

Another way to achieve the same result, but without placing this additional restriction on the tiling procedure itself, is to add wedges and notches to the rhombuses, as illustrated in Figure 5.17.

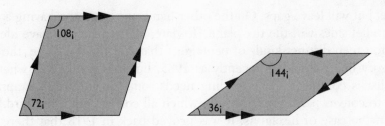

FIGURE **5.16** Penrose tilings. By fitting together the two tiles in such a way that the directions of the arrows match at adjoining edges, it is possible to tile the plane, but only in an aperiodic fashion.

Readers who have by now grown accustomed to the repeated appearance of various numerical patterns in seemingly very different circumstances may not be too surprised to discover that the golden ratio, ϕ (approximately 1.618), is lurking just beneath the surface here. If the sides of the two rhombuses in Figure 5.16 all have length 1, then the long diagonal in the left-hand rhombus is ϕ, and the short diagonal in the right-hand rhombus is $1/\phi$. When the entire plane is tiled using these two figures, the ratio of the number of fat tiles to thin tiles, computed as a limit, is ϕ.

Plate 11 illustrates part of a Penrose tiling of the plane. If you look at it for a while, you will notice that small regions of the tiling pattern have a fivefold symmetry—if you rotate the region as you would a regular pentagon, the region appears to be unchanged. This fivefold symmetry is, however, strictly local. Though various finite regions of the tiling have fivefold symmetry, the entire, infinite tiling does not. Thus, whereas the regular pentagon does not tile the plane, it is nevertheless possible to tile the plane with figures that exhibit local fivefold symmetry. This mathematical discovery, which Penrose made as a result of a

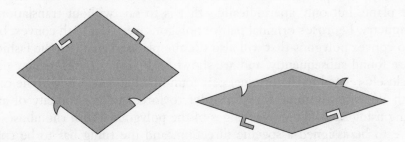

FIGURE **5.17** Penrose tilings. Adding wedges and notches to the tiles gives polygons that will tile the plane—but only aperiodically—without the need for a restriction on the alignment of tiles.

purely recreational investigation, took on added significance in 1984, when crystallographers discovered what are now known as *quasicrystals*.

The crystallographers observed that a certain alloy of aluminum and manganese has a molecular structure that exhibits local fivefold symmetry. Since a crystal lattice can have only two-, three-, four-, or sixfold symmetry, the alloy could not be a crystal in the usual sense, hence the new term 'quasicrystal'. In general, a quasicrystal is a material that, while not having the regular lattice structure of ordinary crystals, nevertheless does have its atoms arranged in a highly ordered fashion that exhibits local symmetry.

Whether the structure of any known quasicrystal is that of the Penrose tilings is not clear, and indeed, the study of quasicrystals is still in its infancy, and not without controversy. Nevertheless, the fact that the plane can be tiled in a highly regular, though non-lattice, fashion that exhibits local fivefold symmetry does demonstrate the possibility of a mathematical framework that can serve as a basis for understanding these newly discovered materials. Once again we have an example of a development in pure mathematics, discovered for its own sake as a mathematician searched for new patterns, preceding a practical application.

Turning to the question of 'tiling' three-dimensional space, it is clear that the cube will completely fill space. In fact, it is the only regular polyhedron that can do so. We have already seen the five nonregular polyhedra that fill space in a lattice fashion, in Figure 5.14.

Three-dimensional analogues of the aperiodic Penrose tilings have been known for several years. Penrose himself found a pair of rhombohedra (squashed cubes) that, when stacked according to various alignment restrictions (much as in the case of the two-dimensional analogue), fill all of three-dimensional space in an aperiodic fashion.

In 1993, to the surprise of most mathematicians, a single convex polyhedron was discovered that could be arranged to fill space completely, but only in an aperiodic fashion. Credit for the discovery of this remarkable solid goes to British mathematician John Horton Conway. Conway's new solid, shown in Figure 5.18, is what is known as a biprism; that is, it is formed from two slant triangular prisms fused together. Its faces consist of four congruent triangles and four congruent parallelograms. In order to fill space using this polyhedron, you proceed layer by layer. Each layer is periodic, but in order to lay a second layer on top of a first, the second must be rotated by a fixed, irrational angle, and this twist ensures that the tiling is aperiodic in the vertical direction. (An aperiodic tiling of space using *nonconvex* polyhedra had previously been achieved in this fashion by Austrian mathematician Peter Schmitt.)

FIGURE **5.18** This photograph shows part of an aperiodic tiling of space using cardboard models of the biprism discovered by John Conway in 1993.

Despite its relevance for designers and its interest for the occasional recreational mathematician, the study of tilings remained a relatively obscure branch of mathematics until the last twenty-five years or so. It is now a thriving area of research that has found a number of surprising applications in other parts of mathematics, as well as to such tasks as the distribution of supplies and the design of circuits. With this increased interest, mathematicians are finding that there is much about tiling that they do not understand, providing still further evidence of the fact that deep and challenging mathematical problems arise in all aspects of our lives.

CHAPTER 6

What Happens When Mathematics Gets into Position

The map that's both right and wrong

The London Underground map shown in Plate 12 was first drawn in 1931. Its creator, Henry C. Beck, was a twenty-nine-year-old temporary draftsman working for the London Underground Group. It took Beck two years of persistent effort to persuade his superiors to accept the now-familiar map for publication. Even then, the Underground publicity department produced the map only in small numbers. They feared that the map's total abandonment of geographic accuracy would render it incomprehensible to the majority of Underground travelers. They were wrong. The public loved it, and by the end of its first year in use, a larger version of the map was posted all over the system. Without the need for any explanation or training, the general public not only coped easily with their first encounter with a genuinely topological representation of the Underground network, but also recognized at once its advantages over the more familiar geometric depictions.

Apart from a number of subsequent additions as the Underground system grew, today's London Underground map remains largely in its original form. Its longevity is a testament to both its utility and its aesthetic appeal. And yet, in geometric terms, it is hopelessly inaccurate. It is certainly not drawn to scale, and if you try to superimpose it on a standard map of London, you will find that the positions of the stations on the Underground map are not at all correct. What *is* correct is the representation of the network: the map tells you what subway line to take to get from point A to point B, and where to change lines if necessary. And this, after all, is the only thing that matters to the Underground traveler—one hardly takes the Underground for the views en route! In this one respect, the Underground map is accurate—completely so. It therefore succeeds in capturing an important pattern in the geography of the London Underground system. That pattern is what mathematicians call a topological pattern.

In two dimensions, the mathematical discipline known as *topology* is sometimes referred to as 'rubber sheet geometry', since it studies the properties of figures that are unchanged by stretching or twisting the surface on which the figures are drawn. The topological nature of the Underground map is illustrated with great frequency in London these days due to the manufacture and sale of souvenir T-shirts bearing the map on the chest. Such maps continue to provide a completely reliable guide to Underground travel, whatever the shape of the body they adorn—though propriety dictates that the student of topology not take this particular aspect of the study too far!

Topology is one of the most fundamental branches of present-day mathematics, having ramifications in many other areas of mathematics as well as in physics and other scientific disciplines. The name *topology* comes from the Greek for 'the study of position'.

The seven bridges of Königsberg

As is so often the case in mathematics, the broad-ranging subject known as topology has its origins in a seemingly simple recreational puzzle, the Königsberg bridges problem.

The city of Königsberg, which was located on the River Pregel in East Prussia, included two islands, joined together by a bridge. As shown in Figure 6.1, one island was also connected to each bank by a single bridge, and the other island had two bridges to each bank. It was the habit of the more energetic citizens of Königsberg to go for a long family walk each Sunday, and, naturally enough, their paths would often take them over several of the bridges. An obvious question was whether there was a route that traversed each bridge exactly once.

Euler solved the problem in 1735. He realized that the exact layout of the islands and bridges is irrelevant. What is important is the way in which the bridges connect—that is to say, the *network* formed by the bridges, illustrated in Figure 6.2. The actual layout of the river, islands, and bridges—that is to say, the geometry of the problem—is irrelevant. Using terms we'll define presently, in Euler's network, the bridges are represented by *edges*, and the two banks and the two islands are repre-

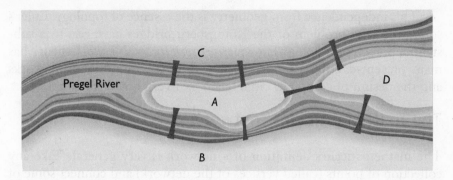

FIGURE **6.1** The Königsberg bridges problem: is it possible to make a tour that traverses each of the seven bridges exactly once? In 1735, Leonhard Euler proved that it is not.

sented by *vertices*. In terms of the network, the problem asks whether there is a path that follows each edge exactly once.

Euler argued as follows: Consider the vertices of the network. Any vertex that is not a starting or a finishing point of such a path must have an even number of edges meeting there, since those edges can be paired off into path-in–path-out pairs. But, in the bridges network, all four vertices have an odd number of edges that meet there. Hence there can be no such path. In consequence, there can be no tour of the bridges of Königsberg that crosses each bridge exactly once.

Euler was able to solve the Königsberg bridges problem by realizing that it had almost nothing to do with geometry. Each island and each bank could be regarded as a single point, and what counted was the way in which these points were connected—not the length or the shape of the connections, but which points were connected to which others.

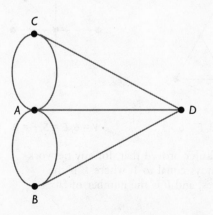

FIGURE **6.2** Euler's solution to the problem of the Königsberg bridges depended on his first observing that its crucial feature was the network formed by the bridges. Points *A, B, C, D*—the vertices—correspond to points *A, B, C, D* in the geographic map in Figure 6.1. The lines—the edges—represent the bridges.

This independence from geometry is the essence of topology. Euler's solution to the problem of the Königsberg bridges gave rise to a substantial branch of topology known as *network theory*. Network theory has many present-day applications; the analysis of communications networks and the design of computer circuits are two very obvious examples.

The mathematicians' network

The mathematician's definition of a network is very general. Take any collection of points (called vertices of the network) and connect some of them together with lines (called edges of the network). The shape of the edges is unimportant, though no two edges may intersect (except at their ends) and no edge may intersect itself or form a closed loop. (You may, however, form a closed circuit by joining together two or more distinct edges at their ends.) In the case of networks on a plane, or on some other two-dimensional surface such as the surface of a sphere, the non-intersection requirement is fairly restrictive. These are the kinds of networks I shall consider here. For networks in three-dimensional space, edges can pass over each other without a problem.

One additional restriction is that a network must be *connected*: that is, it must be possible to go from any vertex to any other vertex by tracing a path along the edges. A number of networks are illustrated in Figure 6.3.

The study of networks in the plane leads to some surprising results. One such is *Euler's formula*, discovered by Euler in 1751. For networks in the plane or on any other two-dimensional surface, the edges of the

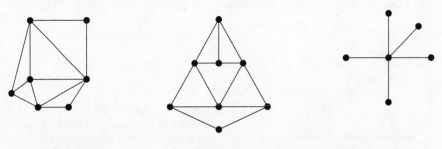

$V = 7, E = 12, F = 6$ $V = 8, E = 13, F = 6$ $V = 6, E = 5, F = 0$

FIGURE **6.3** Euler's formula for networks. Euler proved that, for any network drawn on a plane, the sum $V - E + F$ is always equal to 1, where V is the number of vertices, E is the number of edges, and F is the number of faces enclosed by edges of the network.

network divide the surface into distinct regions, called the *faces* of the network. Take any network and calculate the number of vertices (call that number V), the number of edges (call it E), and the number of faces (call it F). If you now compute the value of the sum

$$V - E + F,$$

you will find that the answer is 1. Always.

This result is clearly quite remarkable: no matter how simple or how complex a network you draw, and no matter how many edges your network has, the above sum always works out to be 1. It is not hard to prove this fact. Here is how it is done.

Given any network, start erasing edges and vertices from the outside, working your way inward, as illustrated in Figure 6.4. Removing one outer edge (but not the vertices at either end) reduces the number E by 1, leaves V unchanged, and decreases F by 1. Thus the net effect on the value of the sum $V - E + F$ is nil, since the reductions in E and F cancel each other out.

Whenever you have a 'dangling edge'—that is, an edge terminating at a vertex to which no other edge is connected—remove that edge and

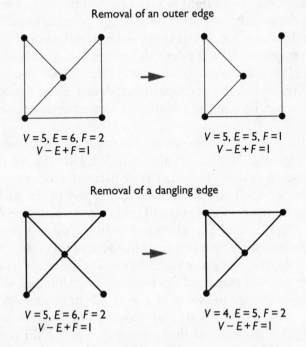

Removal of an outer edge

$V = 5, E = 6, F = 2$
$V - E + F = 1$

$V = 5, E = 5, F = 1$
$V - E + F = 1$

Removal of a dangling edge

$V = 5, E = 6, F = 2$
$V - E + F = 1$

$V = 4, E = 5, F = 2$
$V - E + F = 1$

FIGURE **6.4**   The two key steps in the proof of Euler's formula for networks in a plane.

the free vertex. (If both ends are free, remove one vertex and leave the other as an isolated point. The choice of which to remove and which to leave is up to you.) This reduces both V and E by 1, and leaves F unchanged, so again, the value of the sum $V - E + F$ is not altered by this procedure.

If you continue in this fashion, removing outer edges and pairs of dangling edges and vertices one after another, you will eventually be left with a single, isolated vertex. For this, the simplest of all networks, the sum $V - E + F$ is obviously 1. But the value of this sum does not change at any stage of the reduction procedure, so its value at the end will be exactly the same as at the beginning. So the initial value of $V - E + F$ must have been 1. And there you have your proof!

The fact that the sum $V - E + F$ is equal to 1 for all networks is analogous to, say, the fact that the sum of the angles of any triangle in the plane is 180°. Different triangles can have different angles and sides of different lengths, but the sum of the angles is always 180°; analogously, different networks can have different numbers of vertices and edges, but the sum $V - E + F$ is always 1. But, whereas the triangle angle sum depends crucially on shape, and is thus a fact of geometry, Euler's $V - E + F$ result is completely independent of shape. The lines of the network may be straight or crooked, and the surface on which the network is drawn may be flat or undulating, or even folded. And if the network is drawn on a material that can be stretched or shrunk, neither of these manipulations will affect the result. All of this is surely self-evident. Each of these manipulations will affect geometric facts, but the $V - E + F$ result is not a geometric fact. A fact about figures that does not depend upon bending or twisting or stretching is known as a topological fact.

What happens if the network is drawn not on the plane, but on the surface of a sphere? You can try this by using a felt-tipped pen to draw a network on an orange. Provided your network covers the entire surface (that is, provided you are not using just part of the surface, as if it were simply a curved sheet), you will find that the sum $V - E + F$ turns out to be not 1, but 2. So, whereas bending, twisting, stretching, or shrinking a plane sheet cannot affect the value of this sum, replacing the sheet with a sphere does make a difference. On the other hand, it is easy to convince yourself that bending, twisting, stretching, or shrinking the sphere will not change the value of $V - E + F$ in that case either; it will always be 2. (You can check this by drawing a network on a balloon.)

Proving that the value of the quantity $V - E + F$ for networks on a sphere is always 2 is no more difficult than proving the analogous result for a plane; the same kind of argument can be used. But there is another way to prove this fact. It is a consequence of Euler's formula for networks in the plane, and can be deduced from that previous result by topological reasoning (as opposed to, say, geometric reasoning, which is what Euclid used to prove the theorems in *Elements*).

To begin the proof, imagine your network drawn on a sphere that is perfectly stretchable. (No such material is known, but that's not a problem, since the mathematician's patterns are always in the mind.) Remove one complete face. Now stretch the edges surrounding the missing face so that the entire remaining surface is flattened out into a plane, as illustrated in Figure 6.5. This stretching process clearly will not affect the value of $V - E + F$ for the network, since it will not change any of V, E, or F.

When the stretching is complete, you have a network on the plane. But we already know that, in this case, the value of $V - E + F$ is 1. In going from the original network on the sphere to the network on the plane, all you did was remove one face; the vertices and edges were left intact. So that initial move caused a reduction in the value of $V - E + F$ of 1. Hence, the original value of $V - E + F$ must have been exactly 1 more than the value for the plane network you finished with, which means that the original value for the network drawn on a sphere was 2.

Closely related to this result about networks on spheres is the following fact about polyhedra: If V denotes the number of vertices of a polyhedron, E the number of edges, and F the number of faces, then it is always the case that

$$V - E + F = 2.$$

The topological proof is quite obvious. Imagine that the polyhedron is 'blown up' into the form of a sphere, with lines on the sphere denoting the positions of the former edges of the polyhedron. The result is a network on the sphere, having the same values for V, E, F as the original polyhedron.

In fact, it was in this form, as a result about polyhedra, that the identity $V - E + F = 2$ was first observed by René Descartes in 1639. However, Descartes did not know how to prove such a result. When applied to polyhedra, the result is known as Euler's polyhedra formula.

Möbius and his band

Euler was not the only eighteenth-century mathematician to investigate topological phenomena. Both Cauchy and Gauss realized that figures had properties of form more abstract than the patterns of geometry. But it was Gauss's student Augustus Möbius who really set in motion the

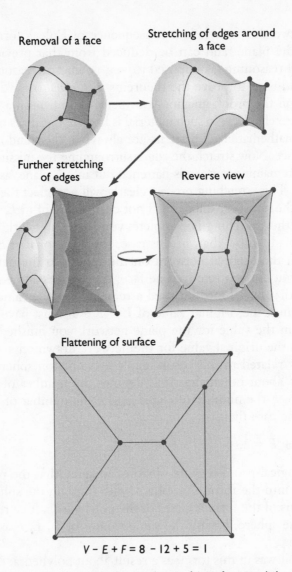

Removal of a face

Stretching of edges around a face

Further stretching of edges

Reverse view

Flattening of surface

$$V - E + F = 8 - 12 + 5 = 1$$

FIGURE **6.5** By removing a single face and stretching the remaining surface to be planar, a network on a sphere can be transformed into a network on a plane that has the same number of vertices and edges, but one less face.

mathematical discipline now known as topology by giving a precise definition of the concept of a topological transformation. According to Möbius's definition, topology is the study of the properties of figures left invariant by topological transformations.

A topological transformation is a transformation from one figure into another such that any two points close together in the original figure remain close together in the transformed figure. It takes a bit of work to make precise just what is meant by the phrase 'close together' in this definition; in particular, stretching is allowed, though this operation clearly increases the distance between points. But the intuition is clear enough. The most significant manipulation that is prohibited by this definition is cutting or tearing, except in the case in which a figure is cut in order to perform some manipulation and then glued back together again so that points on either side of the cut that were originally close together are close together when the transformation is complete.

Most of the early work in topology was directed toward the study of two-dimensional surfaces. A particularly fascinating discovery was made early on by Möbius together with another of Gauss's students by the name of Johann Listing. They found that it was possible to create one-sided surfaces. If you take a long strip of paper—say, one inch wide by ten inches long—give it a single half-twist, and then glue the two free ends together, the result is a surface with only one side, known nowadays as a Möbius band (see Figure 6.6). If you try to color just one side of a Möbius band, you will find that you end up coloring what you might have thought were 'both sides' of the figure.

This, at least, is the way mathematicians often present the Möbius band to children or to beginning students of topology. As is often the case, however, the true story is more subtle.

First of all, *mathematical* surfaces do not have 'sides'. The notion of a side is something that results from observing a surface from the surrounding three-dimensional space. To a two-dimensional creature constrained to live 'in' a surface, the notion of a side makes no sense at all, just as it makes no sense to us to talk about our own, three-dimensional world having sides. (Viewed from four-dimensional space–time, our

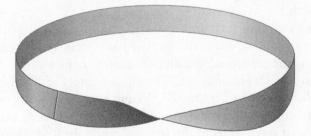

FIGURE **6.6** The Möbius band is constructed by giving a strip of paper a single half-twist and then attaching the two free ends to form a closed loop. It is popularly described as a surface with just one side and one edge.

world does have sides, namely, past and future. But these notions make sense only when you include the additional dimension of time: past and future refer to the position of the world in time.)

Since mathematical surfaces do not have sides, it follows that you cannot draw a figure 'on one side' of a surface. Mathematical surfaces have no thickness; a network may be *in* a surface, but not on it. (Mind you, mathematicians often do speak of "drawing a network on a surface." After all, this terminology is appropriate for the kinds of nonmathematical surfaces we encounter in our daily lives. But when carrying out a mathematical analysis, the mathematician will be careful to treat a network or other figure as being *in* the surface, not 'on' it.)

Thus, it is not mathematically correct to say that the Möbius band has 'one side'. What is it, then, that makes the Möbius band so different from a regular, cylindrical band, formed by looping a strip of paper without giving it an initial half-twist? The answer is that the ordinary band is *orientable*, whereas the Möbius band is *nonorientable*. The mathematical notion of orientability is a genuine property that mathematical surfaces may or may not have. Intuitively, orientability means that there are distinct notions of clockwise and counterclockwise, or that left- and right-handedness are different.

To obtain an initial grasp of this abstract notion, imagine two bands: a simple cylindrical band and a Möbius band, constructed from transparent film, such as clear photographic film or the film used for overhead projector transparencies. (Even better, make two such bands and follow the discussion physically.) Draw a small circle on each band, and add an arrowhead to indicate the clockwise direction. Of course, whether you call this direction clockwise or counterclockwise depends on how you view it from the surrounding three-dimensional space. Using transparent film allows you to see your circle from both 'sides'; the use of this material thus provides a better model of a genuine mathematical surface than does a sheet of opaque paper.

Start with the Möbius band. Imagine sliding your circle all the way around the band until it returns to its original position. To simulate this action on your model, you could start at your circle and proceed around the band in one direction, making successive copies of the circle, together with the directional arrowhead, at some fixed, small distance apart, as illustrated in Figure 6.7. When you have gone completely around the band, you will find yourself making your copy on the other physical side of the film from the original circle. Indeed, you can superimpose the final circle over the original one. But the directional arrowhead, copied from the previous circle, will point in the direction opposite to the original one. In the course of moving around the band, the

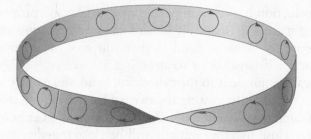

FIGURE **6.7**   The genuine topological property of the Möbius band that corresponds to its 'having only one side' is that it is nonorientable. It is possible to transform a clockwise direction into a counterclockwise direction simply by moving once around the band.

orientation of the circle changes. But since the circle remained 'in' the surface the whole time, and was simply moved around, this result means that, *for this particular surface*, there is no such thing as clockwise or counterclockwise; such a notion simply makes no sense.

On the other hand, if you repeat the procedure with the cylindrical band, the result is quite different. When you complete the circumvention of the band, and the circle has returned to its original position, the directional arrowhead points in the same direction as before. For figures drawn 'in' the cylindrical band, you cannot change the orientation by moving the figure about.

The abstract notion of orientability can also be understood in terms of handedness. If you draw the outline of a human hand on a (transparent) Möbius band and transport it once around the band, you will find that it changes handedness: what you (looking at the band from the surrounding three-dimensional space) perhaps thought of as left- and right-handedness have been interchanged. In the Möbius band, there is no notion of left- or right-handedness.

Orientability is a genuine topological property of surfaces. Since the cylindrical band is orientable and the Möbius band is nonorientable, these two surfaces must be topologically distinct. Hence, it cannot be possible to transform a cylindrical band into a Möbius band by means of a topological transformation. This surely accords with our intuitions. The only way to physically transform a Möbius band into a cylindrical band is by cutting it, removing the half-twist, and then reattaching the two cut ends. But the act of removing the half-twist means that when the two free ends are reattached, points originally close together that were separated by the cut are no longer close together, and thus the transformation is not a topological one.

The distinction between properties of a surface and properties of the surrounding space is dramatically illustrated by constructing a third band, differing from the Möbius band in that you give the strip a full twist (rather than a half-twist) prior to attaching the free ends. This new strip is topologically equivalent to the cylindrical band, since one can be transformed into the other by cutting the twisted band, undoing the twist, and then reattaching the two cut ends. Parts of the band on either side of the cut that were close together initially will be close together after this operation is complete, so this is a genuine topological transformation.

Now perform the following operation on your three bands. Take a pair of scissors and cut along the center, all the way around the band. The result in each of the three cases turns out to be quite different and, if you have not seen it before, quite surprising. In the case of the cylindrical band, you end up with two separate cylindrical bands the same length as the original one. Cutting the Möbius band produces a single, full-twist band, twice as long as the original one. Cutting the full-twist band produces two interlocked full-twist bands, each the same length as the original one. Whereas you might assume that the difference in outcome between cutting the cylindrical band and the Möbius band is attributable to topological differences between these surfaces, the difference in outcomes between cutting the cylindrical band and the full-twist band cannot be explained in this way, since the two are topologically equivalent. The different outcomes arise from the way the bands are embedded in the surrounding three-dimensional space.

You might also try a further experiment. Take each of your three bands, and cut along the length of the band much as before, only start your cut not in the middle of the band, but one-third of the way from one edge. How does the outcome this time compare with the outcome in the previous case?

Orientability is not the only topological property that can be used to distinguish between surfaces. The number of edges is another topological feature of a surface. A sphere has no edges, the Möbius band has one edge, and the cylindrical band has two edges. (You can check that the Möbius band has only one edge by running a colored crayon along the edge.) Thus, the number of edges is another topological property that distinguishes the Möbius band from the cylindrical band. On the other hand, in terms of edges, the Möbius band is the same as a two-dimensional disk, which also has one edge. In this case, orientability is a topological property that distinguishes the two surfaces: the disk is orientable, the Möbius band is not.

What about a disk with a hole somewhere in the middle? This surface is orientable, and has two edges, just like the cylindrical band. And in fact, the two surfaces are topologically equivalent; it is easy to see how to turn a cylindrical band into a disk with a hole in the middle by (mathematical) stretching and flattening out.

All of this is certainly interesting, not to say entertaining. But the topological properties of surfaces are of far more than intrinsic interest. It is generally the case that, when mathematicians discover a truly fundamental kind of pattern, it turns out to have widespread application. This is particularly true of topological patterns.

Historically, what really established the new discipline of topology as a central plank of modern mathematics was the development of complex analysis—the extension of the methods of the differential calculus from the real numbers to the complex numbers, which we explored at the end of Chapter 3.

Since the real numbers lie on a one-dimensional line, a function from real numbers to real numbers can be represented by a line in the plane—a graph of the function. But complex numbers are two-dimensional, and so a function from complex numbers to complex numbers is represented not by a line, but by a surface. The simplest case to imagine is that of a continuous function from complex numbers to real numbers. The graph of such a function is a surface in three-dimensional space, where the real value of the function at any complex point is regarded as a 'height' above or below the complex plane. It was Bernhard Riemann's use of surfaces in complex analysis, around the turn of the century, that brought the study of the topological properties of surfaces to the forefront of mathematics, where it has remained ever since.

How do you tell a coffee cup from a donut?

With the importance of the study of surfaces established, mathematicians needed a reliable means for the topological classification of surfaces. What features of surfaces would suffice to classify surfaces topologically, so that any two surfaces that are topologically equivalent would share those features, and any two surfaces that are not topologically equivalent could be distinguished by one or more of them? In Euclidean geometry, for example, polygons can be classified according to the number of their edges, the lengths of those edges, and the angles between them. An analogous scheme was needed for topology.

A property shared by all topologically equivalent surfaces is called a *topological invariant*. The number of edges is one topological invariant that can be used to classify surfaces. Orientability is another. These suffice to distinguish, say, spheres, cylinders, and Möbius bands. But they do not distinguish between a sphere and a torus (illustrated in Plate 13), both of which have no edges and are orientable. Of course, it is tempting to say that the torus has a hole in the middle and the sphere does not. The problem is that the hole is not a part of the surface itself, any more than sidedness is. The hole of the torus is a feature of the manner in which the surface sits in three-dimensional space. A small, two-dimensional creature constrained to live *in* the surface of the torus would never encounter the hole. The problem for the would-be classifier of surfaces, then, is to find some topological property of the surface that such a creature might be able to recognize, and that is different for the torus and the sphere.

What topological invariants are there, other than number of edges and orientability? One possibility is suggested by Euler's result about the value of $V - E + F$ for networks in a surface. The values of V, E, and F for a given network are unchanged if the surface in which the network is drawn is subjected to a topological transformation. In addition, the value of the quantity $V - E + F$ does not depend on the actual network drawn (at least in the case of the plane or the sphere). So perhaps this quantity $V - E + F$ is a topological invariant of surfaces.

And indeed it is. The kind of reduction argument Euler used to establish the constancy of $V - E + F$ for networks in a plane or in a sphere can be applied to networks in any surface. The constant value of $V - E + F$ for any network in a given surface is known as the *Euler characteristic* of that surface. (You have to make sure that the network genuinely covers the entire surface, and is not just a plane network in a small portion of the surface.) In the case of the torus, the Euler characteristic works out to be 0. Hence this topological invariant serves to distinguish the torus from the sphere, whose Euler characteristic is 2.

Now we have three features to distinguish surfaces: number of edges, orientability, and the Euler characteristic. Are there any others? More to the point, do we need any others, or are these three enough to distinguish any pair of surfaces that are not topologically equivalent?

Perhaps surprisingly, the answer is that these three invariants are indeed all we need. Proving this fact was one of the great achievements of nineteenth-century mathematics.

The key to the proof was the discovery of what are known as *standard surfaces*: particular kinds of surfaces that suffice to characterize all surfaces. It was shown that any surface is topologically equivalent to a sphere having zero or more holes, zero or more 'handles', and zero or more 'crosscaps'. Thus, the topological study of surfaces reduces to an investigation of these modified spheres.

Suppose you have a standard surface that is topologically equivalent to a given surface. The holes in the standard surface correspond to the edges of the original surface. We have seen the simplest case already, when we were verifying Euler's formula for networks on a sphere. We removed one face, creating a hole, and then stretched out the edges of the hole to become the bounding edge of the resulting plane surface. Since this connection between holes in a sphere and the edges of a surface is typical of the general case, from now on I shall restrict attention to surfaces having no edges, such as the sphere or the torus. Such surfaces are called *closed surfaces*.

To attach a *handle* to a surface, you cut two circular holes and sew on a cylindrical tube to join the two new edges, as shown in Figure 6.8. Any closed orientable surface is topologically equivalent to a sphere with a specific number of handles. The number of handles is a topological invariant of the surface, called its *genus*. For each natural number $n \geq 0$, the standard (closed) orientable surface of genus n is a sphere to which are attached n

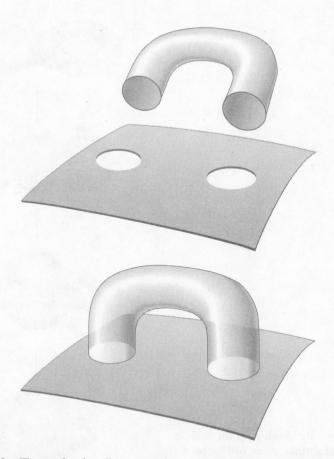

FIGURE **6.8** To attach a handle to a surface, cut two holes in the surface and sew in a hollow, cylindrical tube to join them together.

handles. For example, the sphere is a standard orientable surface of genus 0; the torus is topologically equivalent to a sphere with one handle, the standard orientable surface of genus 1; and the two-holed torus is topologically equivalent to a sphere with two handles, the standard orientable surface of genus 2. Figure 6.9 illustrates a two-holed torus, made from some highly elastic material, that is manipulated to give a sphere with two handles.

The Euler characteristic of a sphere with n handles is $2 - 2n$. To prove this, you start with a (fairly large) network on a sphere (for which $V - E + F = 2$) and then add n handles, one by one. You take care to add each handle so that it connects two holes that result from the removal of two faces of the network. To ensure that the network on the resulting surface genuinely covers the surface, you add two new edges along the handle, as shown in Figure 6.10. Cutting the two holes ini-

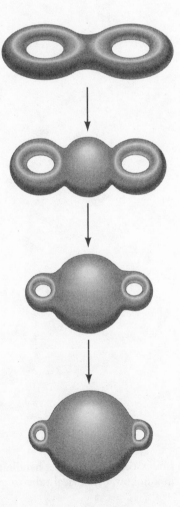

FIGURE **6.9** By blowing up the middle region and shrinking the two loops, a two-holed torus can be transformed into a sphere with two handles.

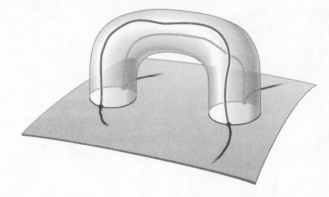

FIGURE **6.10** Adding handles to a sphere in order to calculate the Euler characteristic of the resulting surface.

tially decreases F by 2; sewing on the handle (with its new edges) increases both E and F by 2. The net effect of adding the handle, therefore, is that the quantity $V - E + F$ decreases by 2. This occurs every time you add a new handle, so if n handles are added, the Euler characteristic decreases by $2n$, giving a final value of $2 - 2n$.

The standard nonorientable (closed) surface of genus n is a sphere to which n *crosscaps* have been added. To add a crosscap to a sphere, you cut a hole and sew on a Möbius band, as shown in Figure 6.11. The entire edge of the Möbius band must be sewn to the circular hole; in three-dimensional space, this can be done only if you allow the surface to intersect itself. To sew on the entire edge without such self-intersection, you need to work in four-dimensional space. (Remember, any surface is two-dimensional; the surrounding space is not part of the surface itself. Any surface other than the plane requires at least three dimensions to be constructed. The crosscap is probably the first surface you have encountered that needs four dimensions.)

In order to calculate the Euler characteristic for a sphere with one or more crosscaps, you start with a suitably large network in the sphere, and add the appropriate number of crosscaps, one at a time. Each crosscap replaces one complete face of the network: cut out the face, and sew in a Möbius band along the boundary of that face. Draw a new edge in the Möbius band, connecting the edge to itself, as shown in Figure 6.12. Removing a face reduces F by 1; sewing on the Möbius band adds an edge and a face. Hence the net effect of adding a crosscap is to reduce $V - E + F$ by 1. Thus, the Euler characteristic of a sphere with n crosscaps is $2 - n$.

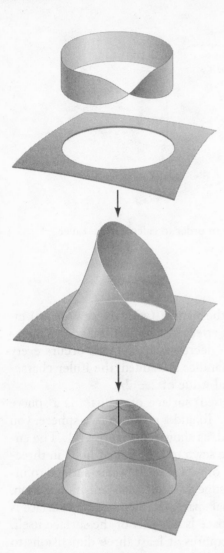

FIGURE **6.11** To add a crosscap to a sphere, cut a hole in the surface and sew in a Möbius band. In three-dimensional space, the attachment of the Möbius band can be achieved only in a theoretical way, by allowing the band to intersect itself. In four dimensions, no such self-intersection is necessary. The crosscap is a surface that requires four dimensions to be constructed properly.

In particular, a sphere with a single crosscap has Euler characteristic $V - E + F = 1$, and a sphere with two crosscaps has Euler characteristic 0. This last surface, which can also be constructed by sewing together two Möbius bands along their single edges, is popularly known as a *Klein bottle*, and is often depicted as in Plate 14. When regarded as some sort of vessel, it has neither inside nor outside. Again, the self-intersection is a result of trying to realize the surface in three-dimensional space; in four dimensions, there is no need for the surface to pass through itself.

In order to complete the classification of all surfaces, it is enough to demonstrate that any closed surface (i.e., any surface without edges) is

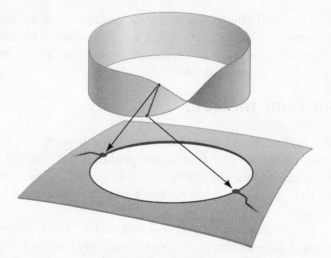

FIGURE **6.12** Adding crosscaps to a sphere in order to calculate the Euler characteristic of the resulting surface.

topologically equivalent to one of the standard surfaces. Given any closed surface, a process of successively removing pieces topologically equivalent to a cylinder or a Möbius band and replacing them with disks leads eventually to a sphere. A cylinder is replaced by two disks, a Möbius band by one. This process is known as surgery. The details are technical, and I will not explain them here.

Finally, what about the title of this section? How do you tell a coffee cup from a donut (Figure 6.13)? If you are a topologist, the answer

FIGURE **6.13** A coffee cup and a donut. To a topologist, the two are identical.

is that you cannot. If you took a plasticine donut, you could manipulate it into the form of a coffee cup (with a single handle). This means that we can describe a topologist as a mathematician who cannot distinguish between a coffee cup and a donut.

The four color theorem

One of the earliest topological questions to be investigated concerned the coloring of maps. The four color problem, formulated in 1852, asked how many colors are needed to draw a map, subject to the requirement that no two regions having a stretch of common border be the same color.

Many simple maps cannot be colored using just three colors. On the other hand, for most maps, such as the county map of Great Britain shown in Plate 15, four colors suffice. The four color conjecture proposed that four colors would suffice to color *any* map in the plane. Over the years, a number of professional mathematicians attempted to prove this conjecture, as did many amateurs. Because the problem asked about all possible maps, not just some particular maps, there was no hope of proving that four colors suffice by looking at any particular map.

The problem is clearly a topological one. What counts is not the shape of the regions of the map, but their configuration—which regions share a common border with which other regions. In particular, the number of colors required will not vary as you manipulate the surface on which the map is drawn—though the answer might vary from one type of surface to another, say, between maps drawn on a sphere and maps drawn on a torus. However, the minimum number of colors required to color any map is the same for maps drawn on a sphere as for maps drawn on a plane, so the four color conjecture may be stated equivalently for maps on a sphere.

In 1976, Kenneth Appel and Wolfgang Haken solved the problem, and the four color conjecture became the four color theorem. A revolutionary aspect of their proof was that it made essential use of a computer. The four color theorem was the first theorem for which no one could read the complete proof. Parts of the argument required the analysis of so many cases that no human could follow them all. Instead, mathematicians had to content themselves with checking the computer program that examined all those cases.

The problem of coloring maps extends naturally to maps drawn on nonplanar surfaces. At the turn of the century, Percy Heawood found a formula that, apart from one exception, seemed to give the minimum number of colors required to color any map on any given closed surface. For a closed surface having Euler characteristic n, the formula predicts the minimum number of colors to be

$$\frac{(7 + \sqrt{49 - 24n})}{2}.$$

For example, according to this formula, the minimum number of colors required to color any map on a torus, for which $n = 0$, is 7. For the sphere, for which $n = 2$, the formula gives the answer 4. (Sadly for Heawood, he was not able to prove that his formula gave the right answer in the case of the sphere, so his formula did not help him to prove the four color conjecture.)

It is now known for certain that Heawood's formula gives the exact minimal number of colors in all cases except for the Klein bottle. For this surface, the Euler characteristic is 0, just as for the torus, so according to the formula, seven colors suffice; but any map on a Klein bottle can be colored using six colors.

Manifold possibilities

A surface can be regarded as being made up of a number—possibly a large number—of small, virtually planar pieces sewn together. Within any one of these component pieces, the surface is just like a piece of the Euclidean plane. The global properties of the surface arise as a result of the manner in which the component pieces are assembled. For example, different means of assembly result in the distinction between a sphere and a torus. On any small region of either, the surface seems like the Euclidean plane, and yet globally, the two surfaces are very different. We are familiar with this phenomenon from our own lives: based purely on our experiences within our everyday, local environments, there is no way we can tell whether the planet on which we live is planar, spherical, or shaped like a torus. This is why the notions and results of Euclidean plane geometry are so relevant to our everyday lives.

A distinction can be made between surfaces that are composed of pieces assembled in a smooth fashion, without any sharp corners or folds, and surfaces assembled with sharp edges, such as polyhedra. Surfaces of the former kind are known as *smooth surfaces*. (Thus, mathematicians assign a technical meaning to the word 'smooth' in this context. Fortunately, this technical meaning accords with the everyday meaning.) In the case of a surface with a sharp edge at some join, the part of the surface surrounding that join does not resemble part of the Euclidean plane.

Based on the ideas discussed above, Riemann introduced the notion of a *manifold* as an important generalization of a surface to higher dimensions. A surface is a two-dimensional manifold, or 2-manifold for short. The sphere and the torus are examples of smooth 2-manifolds. An *n*-dimensional manifold, or *n*-manifold, consists of a number of small pieces joined together, each small piece being, to all intents and purposes, a small region of *n*-dimensional Euclidean space. If the seams where the component pieces are joined together are free of sharp corners or folds, the manifold is said to be smooth.

A fundamental question of physics is, what kind of 3-manifold is the physical universe in which we live? Locally, it looks like three-dimensional Euclidean space, as does any 3-manifold. But what is its global form? Is it everywhere like Euclidean 3-space? Or is it a 3-sphere or a 3-torus, or some other kind of 3-manifold? No one knows the answer.

The nature of the universe aside, the fundamental problem in manifold theory is how to classify all possible manifolds. This means finding topological invariants that can distinguish between manifolds that are not topologically equivalent. Such invariants would be higher-dimensional analogues of orientability and the Euler characteristic, which serve to classify all closed 2-manifolds. This classification problem is by no means solved. Indeed, mathematicians are still trying to overcome an obstacle encountered during the very first investigations into the task of classification.

Henri Poincaré (1854–1912) was one of the first mathematicians to look for topological invariants applicable to higher-dimensional manifolds. In so doing, he helped to found the branch of topology now known as *algebraic topology*, which attempts to use concepts from algebra in order to classify and study manifolds.

One of Poincaré's inventions was what is called the *fundamental group* of a manifold. The basic idea, illustrated in Figure 6.14, is this.

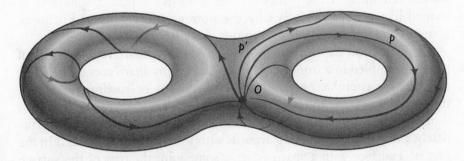

FIGURE **6.14** The fundamental group of a manifold.

You fix some point, O, in the manifold, and consider all loops through the manifold that start and finish at O. Then you try to turn these loops into a group. That means you have to find an operation that can be used to combine any two of these loops into a third, and then verify that this operation satisfies the three axioms for a group. The operation that Poincaré considered is the group sum: if s and t are loops, the group sum $t + s$ is the loop that consists of s followed by t. This operation is associative, so we are already well on the way to having a group. Moreover, there is an obvious identity element: the null loop that never even leaves the point O. So, if every element has an inverse, we will have our group. The next step will be to see to what extent the algebraic properties of the fundamental group characterize the manifold.

There is an obvious candidate for the inverse of a given loop ℓ, namely, the reverse loop, the loop that follows exactly the same path as ℓ, but in the opposite direction. The reverse loop may reasonably be denoted by the symbol $-\ell$. The problem is, although $-\ell$ undoes the effect of ℓ, the combination $-\ell + \ell$ is not the null loop, just as flying from New York to San Francisco and then flying back again is not the same as never leaving New York in the first place. True, in both cases you start and finish in New York, but what happens in between is quite different.

The way out of this dilemma is to declare any two loops to be identical if one may be continuously deformed into the other within the manifold. For example, in Figure 6.14, the path p may be continuously deformed (i.e., topologically transformed) into p' within the manifold. This does the trick, since $-\ell + \ell$ can clearly be transformed continuously into the null loop.

A continuous transformation from one loop or path to another is known as a *homotopy*, and Poincaré's fundamental group, obtained in this manner, is known as a *homotopy group* of the manifold. In the highly simplified case in which the manifold is a circle (a 1-manifold), the only difference between two loops is the number of times each winds around the circle, and the fundamental group in this case turns out to be the group of the integers under addition.

On their own, fundamental groups do not suffice to classify manifolds. But the general idea was a good one, so Poincaré and other mathematicians took it further. By using n-dimensional spheres instead of one-dimensional loops, for each natural number n, they constructed what is known as a (*higher*) *homotopy group* of dimension n. Any two topologically equivalent manifolds must have the same homotopy groups. The question then was, does the collection of all homotopy groups suffice to distinguish any two manifolds that are not topologically equivalent?

With the classification problem for 2-manifolds already solved, the first case to look at was dimension 3. A special instance of that case was the question, if a 3-manifold $\mathcal{M}$ has the same homotopy groups as the 3-sphere, $\mathcal{S}^3$, is $\mathcal{M}$ topologically equivalent to $\mathcal{S}^3$? It was Poincaré himself who first raised this question, in 1904, and the conjecture that the answer is yes became known as the *Poincaré conjecture*.

The Poincaré conjecture generalizes to the case of *n*-manifolds in a straightforward manner: If an *n*-manifold $\mathcal{M}$ has the same homotopy groups as the *n*-sphere, $\mathcal{S}^n$, is $\mathcal{M}$ topologically equivalent to $\mathcal{S}^n$?

Using the classification of 2-manifolds, the answer may be shown to be yes in the case $n = 2$. (The question reduces to looking at homotopy groups associated with the standard surfaces.) But for many years, no one was able to make much headway with any of the higher-dimensional cases, and the Poincaré conjecture began to achieve the same status in topology that Fermat's last theorem enjoyed in number theory. In fact, this comparison does not really do full justice to the Poincaré conjecture. Whereas Fermat's last theorem grew more famous the longer it remained unproved, it did not have any major consequences. The Poincaré conjecture, in contrast, is the key to a whole new area of mathematics, a fundamental obstacle that stands in the way of further progress in our understanding of manifolds.

Given the relative ease with which the two-dimensional case of the Poincaré conjecture can be proved, one might imagine that the next to be proved would be dimension 3, then dimension 4, and so forth. But issues of dimension do not always work that way. Though the complexity and difficulty of problems generally does increase as you go up through dimensions 1, 2, and 3, it can happen that things simplify enormously when you get to dimension 4 or 5 or so. The additional dimensions seem to give you more room to move about, more scope to develop tools to solve your problem.

This is exactly what happened in the case of the Poincaré conjecture. In 1961, Stephen Smale proved the conjecture for all dimensions *n* from $n = 7$ upward. Soon afterward, John Stallings pushed the result down to include the case $n = 6$, and Christopher Zeeman took it a step further to include $n = 5$. Only two more cases were left to be proved!

A year went by, then two, then five. Then a decade. Then two decades. Further progress began to seem hopeless.

Finally, in 1982, Michael Freedman broke the deadlock, finding a way to establish the truth of the Poincaré conjecture for dimension 4. That left only the case $n = 3$. And there, to everyone's frustration, the matter remains to this day. With all cases of the Poincaré conjecture verified apart from dimension 3, it is tempting to assume that the conjecture is indeed true for all dimensions. And most topologists probably expect this to be the case. But expectation is not proof, and for the moment, the Poincaré conjecture remains one of the greatest unsolved problems in topology.

One possible way of approaching the problem—and indeed, the overall classification of 3-manifolds—is to make use of techniques from geometry. This, at least, was the proposal put forward by the mathematician William Thurston during the 1970s. Thurston's approach is reminiscent of Klein's Erlangen program, in which group theory was used to study geometry (see page 198). Despite the fact that topological properties are highly nongeometric, Thurston thought that geometric patterns might nevertheless prove useful for the study of 3-manifolds.

Such a program is not easily carried out. For one thing, in three dimensions, there are eight different geometries to deal with, as Thurston himself proved in 1983. Three of these correspond to the three planar geometries: namely, Euclidean 3-space, elliptic 3-space (which corresponds to two-dimensional Riemannian geometry), and hyperbolic 3-space (which corresponds to two-dimensional hyperbolic geometry). The remaining five geometries are new ones, which arose as a result of Thurston's investigation.

Though Thurston's program is by no means complete, considerable progress has been made, demonstrating once again the incredible power of cross-fertilization in mathematics when patterns from one area are applied to another. In this case, Thurston's program analyzes the group-theoretic patterns of possible geometries (remember, any geometry is essentially determined by a particular group of transformations) and then applies those geometric patterns to the topological study of 3-manifolds.

It should be said that the expectation that the one remaining case of the Poincaré conjecture will turn out to be true is certainly not based on the fact that all the other cases have been proved. If ever topologists had thought that all dimensions behaved in more or less the same way, they were forced to change their views radically by an unexpected, and dramatic, discovery made in 1983, by a young English mathematician named Simon Donaldson.

Physicists and engineers make frequent use of the differential calculus in the study of Euclidean space of dimension 3 or more. Generalizing the notion of differentiation from the two-dimensional case described in Chapter 3 is relatively straightforward. However, physicists, in particular, need to be able to use techniques of the differential calculus on smooth manifolds other than Euclidean n-space. Since any n-manifold can be split up into small pieces that each look like Euclidean n-space, for which we know how to do differentiation, it is possible to apply the differential calculus to a manifold in a strictly local way. The question is, can differentiation be carried out globally? A global scheme for differentiation is called a *differentiation structure*.

By the mid-1950s it was known that any smooth 2- or 3-manifold could be given a unique differentiation structure, and it was assumed that this result would eventually be extended to higher dimensions. But, to everyone's surprise, in 1956 John Milnor discovered that the 7-sphere could be given 28 distinct differentiation structures, and soon afterward similar results were discovered for spheres of other dimensions.

Still, topologists could comfort themselves that these new results did not apply to Euclidean n-space itself. It was surely the case, they thought, that there was only one differentiation structure for these familiar spaces, the spaces to which the original methods of Newton and Leibniz apply.

Or was it? Certainly, it was known that both the Euclidean plane and Euclidean 3-space have a unique differentiation structure, namely, the standard one. It was also known that the standard differentiation structure is unique for Euclidean n-space for all values of n other than 4. Curiously, however, no one had been able to give a proof for the four-dimensional case. But it was surely just a matter of time before someone hit upon the right combination of ideas, wasn't it? The mathematicians were all the more frustrated at their seeming inability to provide a proof because this was the very space of most concern to physicists working in four-dimensional space–time. The physicists were waiting for their colleagues, the mathematicians, to resolve the matter.

But this turned out to be a rare occasion when the usual way of doing business was reversed. Instead of the physicists making use of new ideas from mathematics, the methods of physics came to the rescue of the mathematicians. In 1983, using ideas from physics known as Yang–Mills gauge fields, which had been introduced in order to study the quantum behavior of elementary particles, together with the methods Michael Freedman had developed in order to prove the Poincaré conjecture in dimension 4, Simon Donaldson showed how to construct a nonstandard differentiation structure for Euclidean 4-space, in addition to the usual one.

In fact, the situation rapidly became even more bizarre. Subsequent work by Clifford Taubes demonstrated that the usual differentiation structure for Euclidean 4-space is just one of an infinite family of different differentiation structures! The results of Donaldson and Taubes were completely unexpected, and quite contrary to everyone's intuitions. It seems that not only is Euclidean 4-space of particular interest because it is the one in which we live (if you include time), it is also the most interesting—and the most challenging—for the mathematician.

We shall pick up this work again presently. In the meantime, we'll take a look at another branch of topology: the study of knots.

Mathematicians tie themselves up in knots

The first book on topology was *Vorstudien zur Topologie*, written by Gauss's student Listing and published in 1847. A considerable part of this monograph was devoted to the study of knots, a topic that has intrigued topologists ever since.

Figure 6.15 illustrates two typical knots, the familiar overhand knot and the figure-of-eight. Most people would agree that these are two different knots. But what exactly does it mean for two knots to be different? Not that they are made from different pieces of string; nor does the actual shape of the string matter. If you were to tighten either of these knots, or change the size or shape of the loops, its overall appearance would change, perhaps dramatically, but it would still be the same knot. No amount of tightening or loosening or rearranging would seem to turn an overhand knot into a figure-of-eight knot. The distinctive feature of a knot, surely, is its knottedness, the manner in which it loops around itself. It is this abstract pattern that mathematicians set out to study when they do knot theory.

Since the knottedness of a string does not change when you tighten it, loosen it, or manipulate the shape of the individual loops, knot patterns are topological. You can expect to make use of some of the ideas

The overhand

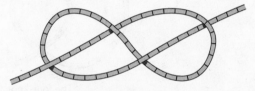

The figure-of-eight

FIGURE **6.15**   Two familiar knots: the overhand and the figure-of-eight.

and methods of topology in studying knots. But you have to be careful. For one thing, there is surely a very easy way to manipulate, topologically, the overhand knot so that it becomes a figure-of-eight knot: simply untie the overhand knot and retie it as a figure-of-eight. There is no cutting or tearing involved in this process; topologically speaking, the procedure is perfectly in order. But, obviously, if you want to make a mathematical study of knots, you want to exclude the case of transforming one knot into another by untying the one and then retying it in the form of the other.

So, when mathematicians study knots, they require that the knot have no free ends, as shown in Figure 6.16. Before analyzing a particular knot, the mathematician first joins together the two free ends to form a closed loop. The two knots illustrated result from joining the free ends of the overhand and figure-of-eight; they are called the *trefoil* and the *four-knot*, respectively. In the case of physical knots, made out of string, attaching the two ends would mean gluing them together.

Restricting our attention to knots tied in closed loops of string solves the problem of trivializing the study by being able to untie and retie, and yet clearly retains the essential notion of knottedness. It is surely impossible to transform the trefoil knot shown in Figure 6.16 into the

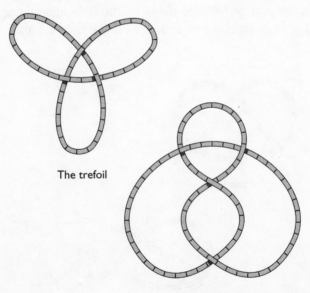

The trefoil

The four-knot

FIGURE **6.16** Two mathematical knots: the trefoil and the four-knot. Mathematical knots consist of closed loops in space.

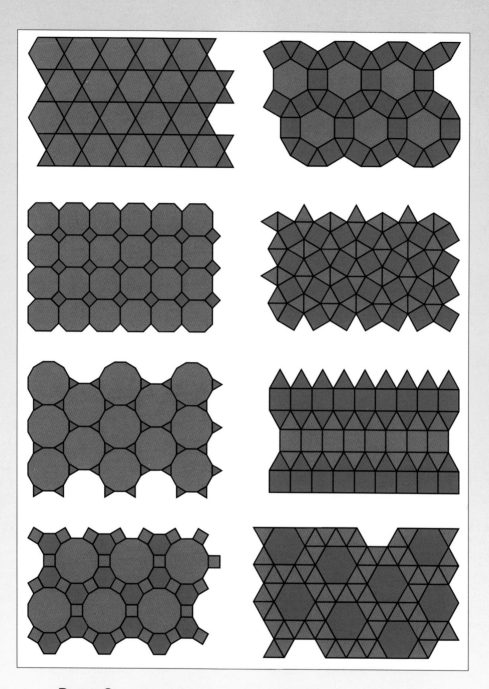

PLATE 9   The eight ways to tile the plane with two or more regular polygons, when it is required that the arrangement of polygons around each vertex be the same.

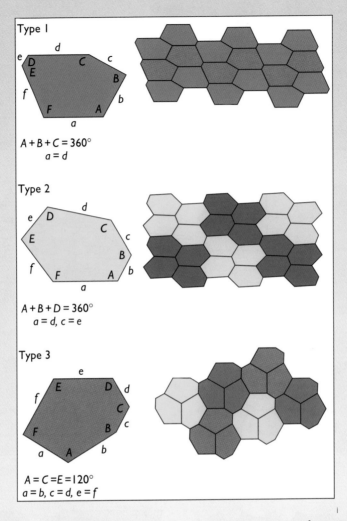

Type 1

$A + B + C = 360°$
$a = d$

Type 2

$A + B + D = 360°$
$a = d, c = e$

Type 3

$A = C = E = 120°$
$a = b, c = d, e = f$

PLATE 10   The three ways to tile the plane using a convex hexagon. In order for a hexagon to tile the plane, it must satisfy certain conditions, which are stipulated by the equations given with each figure. For example, in the first figure, the angles A, B, and C must sum to 360° and sides a and d must be equal in length. The coloring indicates the basic tiling pattern, which is repeated indefinitely by translation to tile the entire plane.

PLATE 11   Part of an (aperiodic) Penrose tiling of the plane,
exhibiting the local fivefold symmetry of the tiling.

PLATE 12   Map of the London Underground.

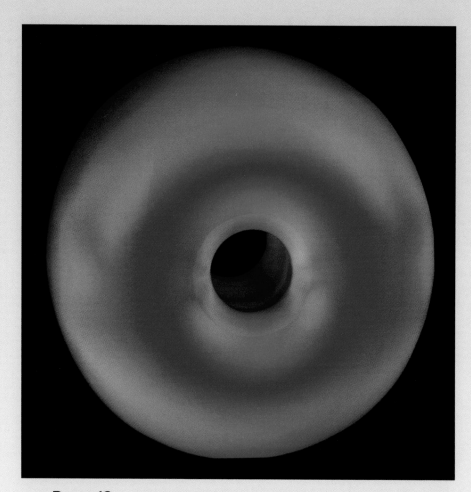

PLATE 13 A computer-generated image of a torus.

PLATE 14   The Klein bottle. Popularly described as a container having no 'inside' or 'outside', this surface can be constructed
in three-dimensional space only if you allow the surface to pass
through itself. In four dimensions, it is possible to construct a
Klein bottle without the need for such self-intersection.
Topologically, you can obtain a Klein bottle by sewing together
two Möbius bands along their (single) edges.

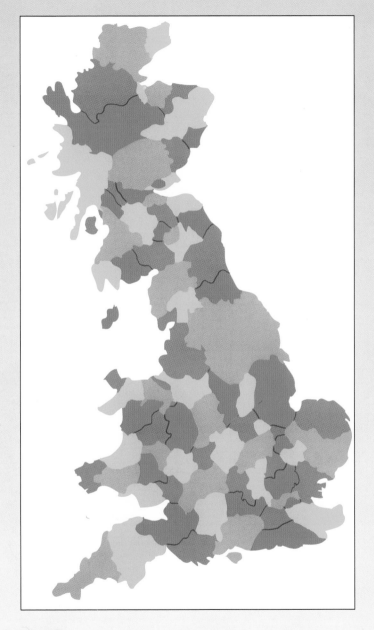

PLATE 15 A county map of Great Britain can be drawn using just four colors.

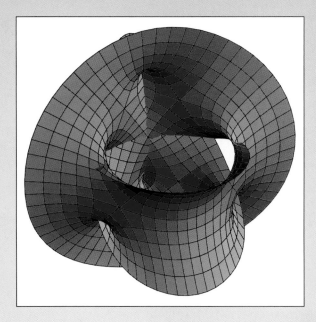

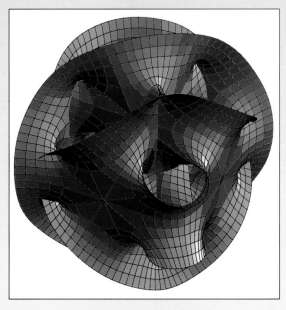

PLATE 16   Surfaces generated by the Fermat equation
$x^n + y^n = 1$, where x and y are regarded as complex variables. The
figure on the left shows the surface for $n = 3$; the one on the right
is for $n = 5$. Both figures were produced by the mathematics
software system *Mathematica*, one of a number of sophisticated
computer tools available to the modern mathematician.

four-knot shown in the same figure. (Try it for yourself and see. Construct a trefoil out of string, join the ends together, and then attempt to turn your knot into a four-knot, without undoing the two ends.)

Having decided to ignore the material a knot is constructed from, and to insist that a knot has no free ends, we arrive at the mathematician's definition of a knot: it is a closed loop in three-dimensional space. (From this standpoint, the two 'knots' shown in Figure 6.15 are not knots at all.) As a loop in space, a mathematical knot has no thickness, of course; it is a one-dimensional entity—a 1-manifold, to be precise.

Our task now is to study the patterns of knots. This means ignoring issues of tightness, size, the shape of individual loops, and the position and orientation of the knot in space; mathematicians do not distinguish between knots that are topologically equivalent.

But just what is meant by that last phrase, 'topologically equivalent'? There is still a very simple way to transform, say, a trefoil into a four-knot in a topologically legitimate fashion: cut the string, untie it, retie it in the form of a four-knot, and then fasten the free ends back together again. Points close together before this process remain close together afterward, so this is a permissible topological transformation. But it clearly violates the spirit of what we are trying to do: the one thing we do not want to allow in the study of knots is cutting.

The point about a mathematical knot is that its pattern arises from the manner in which it is situated in the surrounding three-dimensional space. That pattern is topological, but the topological transformations that are relevant are transformations of all of 3-space, not just the knot. When mathematicians speak of two knots being topologically equivalent (and hence, in fact, the 'same' knot), they mean that there is some topological transformation of 3-space that transforms one knot into the other.

Though this official definition of knot equivalence is important in the detailed mathematical study of knots, it is not very intuitive. But the essence of the definition is that it excludes cutting the knot loop, and yet allows any other topological manipulation. Indeed, it is only in more advanced work in knot theory that topologists look closely at the whole of space, as they examine the complicated 3-manifold that remains when a knot is removed from 3-space.

The study of knots is a classic example of the way in which mathematicians approach a new area of study. First, a certain phenomenon is observed—in this case, knottedness. Then the mathematician abstracts away from all those issues that appear irrelevant to the study, and formulates precise definitions of the crucial notions. In this case, those crucial notions are knots and knot equivalence. The next step is to find ways to describe and analyze the different kinds of knots—the different knot patterns.

For example, what distinguishes the trefoil knot from the null knot—an unknotted loop? Of course, they look different. But, as mentioned earlier, what a knot looks like—that is to say, the manner in which a particular knot is laid out, or *presented*—is not important. The question is, can the trefoil be manipulated into an unknotted loop without cutting the loop? It certainly looks as though it cannot. Moreover, if you were to make a trefoil out of string and play around with it for a while, you might find you were unable to unknot it. But this does not amount to a proof. Maybe you simply have not tried the right combination of moves.

(Actually, it is arguable that, in the case of a very simple example such as the trefoil, mental or physical manipulation does amount to a proof, in all but the strictest sense of formal logic—a standard of proof that almost no real theorem of mathematics ever adheres to. But in the case of more complicated knots, such an approach would not constitute a proof. Besides, what is wanted is not a method for coping with one particular, very simple example, but a general method that works for all knots, including those nobody has seen yet. One should always be cautious in using examples. To serve as such, an example has to be simple, but its purpose is to help us understand the underlying issues that apply to more complicated cases.)

A more reliable method of distinguishing two knots would be to find some knot invariant for which they differ, a *knot invariant* being any property of a knot that does not change when you subject the knot to any permissible manipulation. In order to look for knot invariants, you must first find some way to represent knots. An algebraic notation might help eventually, but at the outset of the study, the most obvious representation is a diagram. Indeed, I have already presented two knot diagrams in Figure 6.16. The only modification that mathematicians make when they draw such diagrams is that they do not try to draw a picture of a physical knot, made from string or rope, but use a simple line drawing that indicates the knot pattern itself. Figure 6.17 gives some examples, including the mathematician's version of the trefoil, illustrated previously in Figure 6.16. The lines are broken to indicate where the knot passes over itself. Such a diagrammatic representation of a knot is often referred to as a *presentation* of the knot.

One way to understand the structure of a complicated knot is to try to split it up into a number of smaller, simpler knots. For example, both the reef knot and the granny knot can be split up into two trefoils. Expressing this the other way round, one way to construct a (mathematical) reef or granny is to tie two trefoils on the same mathematical string,

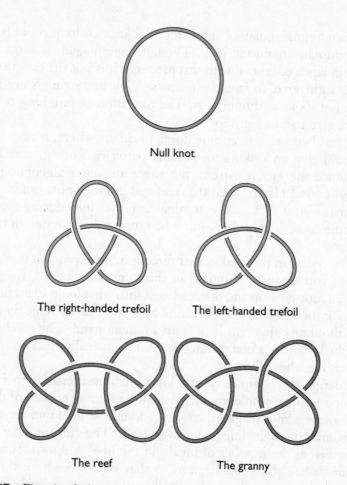

Null knot

The right-handed trefoil   The left-handed trefoil

The reef   The granny

FIGURE **6.17** *Five simple knots, as a knot theorist would typically represent them. After a few moments' consideration, it seems obvious—though it is not easy to prove—that each of these five knots is different from all the others; that is to say, no amount of manipulation of any one of these knots will transform it into one of the others.*

and then join the free ends. It is natural to describe this process of tying two knots on the same string as constructing the 'sum' of the two knots.

This summation operation is associative, and the null knot is obviously an identity operation. At this point the mathematician, ever on the lookout for new patterns, will begin to wonder if this is yet another example of a group. All that would be required is for every knot to have an inverse. Given any knot, is it possible to introduce another knot on the same mathematical string so that the result is the null knot? If it were, then by manipulating the string, the knotted loop could be transformed into an unknotted loop. Though stage magicians know how to tie certain kinds of 'knots' with this property, it is not the case that every knot has an inverse. In fact, the magician's knotted string is not knotted at all; it just looks as though it is. The operation of summing two knots does not give rise to a group.

But just because one avenue doesn't lead anywhere, it doesn't mean you should give up looking for familiar patterns. Knot summation may not produce the group pattern, but there are other algebraic patterns that might arise. Having seen that reef and granny knots can be split up into sums of simpler knots, you might consider introducing the notion of a 'prime' knot, a knot that cannot be expressed as the sum of two simpler knots.

Before you can proceed in that direction, however, you ought to say just what you mean by 'simpler' in this context. After all, it is easy to take an unknotted loop and manipulate it into what looks like a fiendishly complex knot—fine necklace chains have a habit of achieving such a state without any apparent help from a human hand. The resulting tangle might look like a complicated knot, but in reality it is the simplest knot possible, the null knot.

To define the complexity of a knot, mathematicians associate with any knot a positive whole number, called the *crossing number* of the knot. If you look at a knot diagram, you can add up the total number of crossings—points where the line passes over itself. (The same number can also be described as the number of breaks in the line, as drawn. You have to first manipulate the knot diagram so that you never have three lines crossing at the same point.) The number of crossings provides you with a measure of the complexity of the diagram. Unfortunately, it tells you little about the actual knot. The problem is, the same knot can have infinitely many different numbers associated with it in this way: you can always increase the number by 1 as many times as you like, without changing the knot, by the simple act of introducing new twists in the loop.

But, for any knot, there will be a unique *minimum* number of crossings you can obtain in this way. That minimum number clearly is a measure of the knot's complexity; it is the number of crossings in a diagram that represents the knot in the simplest possible way, devoid of any superfluous twists in the loop. It is this minimum number of crossings that is called the crossing number. It tells you how many times the loop is forced to cross itself in order to produce the knot, regardless of how many times it actually does cross itself in a particular presentation of the knot. For example, the crossing number of the trefoil is 3, and that of both the reef and the granny is 6.

You now have a way to compare two knots: knot A is simpler than knot B if A has a smaller crossing number than B. You can proceed to define a prime knot as one that cannot be expressed as the sum of two simpler knots (neither of which is the null knot).

A great deal of the early work in knot theory consisted of attempts to identify all the prime knots with a given crossing number. By the turn of the century, many prime knots with crossing numbers up to 10 had been identified; the results were presented as tables of knot diagrams (see Figure 6.18).

The work was fiendishly difficult. For one thing, in all but the simplest cases, it is extremely hard to tell whether two different-looking di-

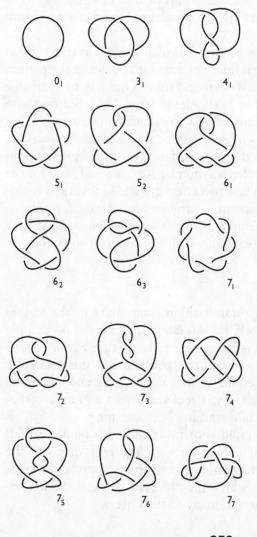

FIGURE **6.18** A table of knot diagrams showing all knots with seven or fewer crossings.

agrams represent the same knot, so no one was ever sure if the latest table had duplicate entries. Still, with the work of J. W. Alexander and G. B. Briggs in 1927, mathematicians knew there were no duplications in the tables for crossing numbers as far as 8, and, a short while later, H. Reidemeister sewed up the case for crossing number 9 as well. The case of crossing number 10 was finally settled in 1974, by K. A. Perko. All of this progress was achieved by using various knot invariants to distinguish between different knots.

The crossing number is a very crude knot invariant. Though knowing that two knots have different crossing numbers does tell you that the knots are definitely not equivalent, and, indeed, provides a way to compare the complexities of the knots, there are simply too many knots having the same crossing number for this means of knot classification to be of much use. For instance, there are 165 prime knots with crossing number 10.

One reason why the crossing number is such a weak invariant is that it simply counts the crossings; it does not attempt to capture the pattern of the crossings as the knot weaves around itself. One way to overcome this deficiency was discovered by J. W. Alexander in 1928. Starting with the knot diagram, Alexander showed how to calculate not a number, but an algebraic polynomial, known nowadays as the *Alexander polynomial*. The exact details of how to do this are not important in the present context; for the trefoil, the Alexander polynomial is $x^2 - x + 1$, and for the four-knot it is $x^2 - 3x + 1$. When two knots are added together, multiplying the two Alexander polynomials produces the Alexander polynomial of the sum knot. For example, the Alexander polynomial of both the reef and the granny knot, each of which is a sum of two trefoils, is

$$(x^2 - x + 1)^2 = x^4 - x^3 + 3x^2 - 2x + 1.$$

Because it captures, in an algebraic fashion, something of the way in which a knot winds around itself, the Alexander polynomial is certainly a useful knot invariant. And it is fascinating to know that a knot pattern can be partially captured by an algebraic pattern. But the Alexander polynomial is still somewhat crude; it does not capture enough of the pattern of a knot to distinguish, say, a reef knot from a granny, something done with ease by any child who has been camping.

Other attempts to find simple knot invariants also tended to fall short when it came to distinguishing between the reef knot and the granny. But for all their shortcomings, these various approaches highlight yet again the manner in which the different patterns the mathematician weaves find application in many distinct areas.

The Alexander polynomials are derived from another, fairly powerful knot invariant called the *knot group*. This is the fundamental group (or homotopy group) of the *knot complement*, the 3-manifold that is left when the knot itself is removed. The members of this group are closed, directed loops that start and finish at some fixed point not on the knot, and which wind around the knot. Two loops in the knot group are declared identical if one can be transformed into the other by manipulating it in a manner that does not involve cutting or passing through the knot. Figure 6.19 illustrates the knot group for the trefoil.

The knot group provides the mathematician with a means of classifying knots according to properties of groups. An algebraic description of the knot group can be derived from the knot diagram.

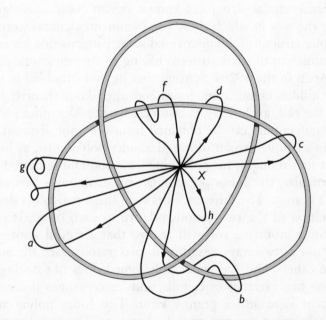

FIGURE **6.19** The knot group for the trefoil. The members of the group are closed, directed loops that start and finish at point X. Loops *a*, *b*, and *g* are regarded as the same, since one can be transformed into the other without cutting or passing through the knot. Loops *c* and *d* are regarded as distinct, since they pass around the knot in opposite directions. Loop *h* is identified with the null loop of zero length, the identity element of the group. The group operation is the combination of loops, with the 'sum' $x + y$ of loops x and y consisting of loop y followed by loop x. (The intermediate step of passing through the basepoint X is ignored when two loops are combined.) For example, $d + d = f$ and $c + d = h$.

Another ingenious way of classifying knots is to construct, for a given knot, an orientable (i.e., 'two-sided') surface having the knot as its only edge, and to take the genus of that surface as an invariant of the knot. Since there may be more than one surface that can be associated with the same knot in this fashion, you take the smallest genus number that can arise in this way. The resulting number, called the genus of the knot, is a knot invariant.

However, none of the knot invariants mentioned so far can distinguish a reef knot from a granny, and for many years there seemed to be no hope of finding a simple way to do this. (Knot theorists could make the distinction using more complex ideas.) But all of that changed in 1984, when a New Zealand mathematician by the name of Vaughan Jones discovered a quite new class of polynomial invariants for knots.

The discovery was made quite by chance. Jones was working on a problem in analysis that had applications in physics. The problem concerned mathematical structures known as von Neumann algebras. In looking at the way in which these von Neumann algebras were built up from simpler structures, he discovered some patterns that his colleagues found reminiscent of some patterns having to do with knots, discovered by Emil Artin in the 1920s. Sensing that he had stumbled onto an unexpected, hidden connection, Jones consulted knot theorist Joan Birman, and the rest, as they say, is history. Like the Alexander polynomial, the Jones polynomial can be obtained from the knot diagram. But, far from being a simple variant of the Alexander polynomial, as Jones himself had at first thought, his polynomial was something quite new.

In particular, the Jones polynomial can distinguish between a reef knot and a granny. The difference between these two knots depends on the orientation of the two trefoils relative to each other. If you think about it for a moment, you will realize that a trefoil knot can wrap around in one of two ways, resulting in two trefoils, each the mirror image of the other. The Alexander polynomial does not distinguish between these two varieties of trefoils, and hence cannot distinguish between a reef knot and a granny knot. The Jones polynomial does distinguish between the two trefoils, however. The two Jones polynomials are:

$$x + x^3 - x^4$$
$$x^{-1} + x^{-3} - x^{-4}$$

Strictly speaking, the second of these is not a polynomial, since it contains negative powers of the variable x. But in this case, mathematicians use the word anyway.

In fact, not only was Jones's initial breakthrough significant in itself, it also opened the way to a whole array of new polynomial invariants, and led to a dramatic rise in research in knot theory, some of it spurred on by the growing awareness of exciting new applications in both biology and physics, both of which we consider briefly.

First, biology. A single strand of human DNA can be as long as 1 meter. Coiled up, it can fit into a cell nucleus having a diameter of about five-millionths of a meter. Clearly, the DNA molecule has to be pretty tightly interwoven. And yet, when the DNA divides to give two identical copies of itself, these two copies slide apart in an effortless way. What kind of knotting permits this smooth separation to happen? This is just one of many questions that face biologists in their quest to understand the secrets of life itself.

It is a question that might be answered with the help of mathematics. Since the mid-1980s, biologists have teamed up with knot theorists in an attempt to understand the knot patterns nature uses in order to store information in genes. By isolating single DNA strands, fusing their ends to create a mathematical knot, and then examining them under a microscope, it has been possible to apply mathematical methods, including the Jones polynomials, in order to classify and analyze these fundamental patterns (see Figure 6.20).

One important application of this work has been in finding ways to combat viral infections. When a virus attacks a cell, it often changes the knot structure of the cell's DNA. By studying the knot structures of the infected cell's DNA, researchers hope to gain an understanding of the way the virus works, and thereby develop a countermeasure or a cure.

Turning from biology to physics, we should observe that as early as 1867, Lord Kelvin put forward an atomic theory proposing that atoms

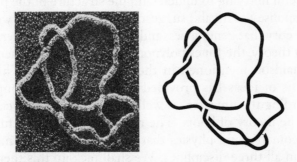

FIGURE **6.20** *Left:* An electron microscope photo of a strand of DNA. *Right:* A line drawing showing the knot structure formed by the DNA molecule.

were knots in the ether. Known as the theory of vortex atoms, this proposal had some sensible reasons behind it. It explained the stability of matter, and it provided for a large collection of different atoms, taken from the rich collection of knots then in the process of being classified by knot theorists. It also provided an explanation of various other atomic phenomena. Kelvin's theory was taken sufficiently seriously to provide the impetus for some of the early mathematical work on the classification of knots; in particular, his collaborator P. G. Tait produced extensive knot tables. But for all its mathematical elegance, the theory of vortex atoms went the same way as Plato's atomic theory, described in Chapter 4. It was eventually replaced by the idea put forward by Niels Bohr, of the atom as a miniature solar system.

These days, with Bohr's theory having been abandoned as too naive, knot theory has once again come to the fore. Physicists now suggest that matter is made up of so-called superstrings—tiny, knotted, closed loops in space–time, whose properties are closely bound up with their degree of knottedness.

To return to our main theme: in 1987, after the discovery of the Jones polynomials, further polynomial invariants for knots were found that were based on ideas from statistical mechanics, an area of applied mathematics that studies the molecular behavior of liquids and gases. Not long afterward, it was observed that the knot-theoretic pattern captured by the Jones polynomial itself arises in statistical mechanics. Knots, it seems, are everywhere—or, more precisely, the patterns that knots exhibit are everywhere.

The ubiquity of knots was illustrated in a particularly dramatic and far-reaching fashion by the rapid rise of topological quantum field theory, a new theory of physics developed by Edward Witten in the late 1980s. The mathematical physicist Sir Michael Atiyah was the first to suggest that the mathematical patterns captured by the Jones polynomial might be useful in trying to understand the structure of the physical universe. In response to Atiyah's suggestion, Witten came up with a single, very deep theory that generalizes, and builds upon, the patterns captured by quantum theory, the Jones polynomials, and the fundamental work of Simon Donaldson mentioned in the previous section. This powerful new synthesis of ideas has provided physicists with a completely new way of regarding the universe, as well as giving mathematicians new insights into the theory of knots. The result has been a fruitful merger of topology, geometry, and physics that promises to lead to many further discoveries in all three disciplines. We shall pick up this theme again in Chapter 8. In the meantime, we should observe that, in developing the mathematical theory of knots, mathematicians have created new ways of understanding certain aspects of the world, both the living world of DNA and the physical universe we live in. After all, what is understanding other than a recognition of a pattern of some kind or another?

Fermat's last theorem again

Now, at last, it is possible to complete our account of Fermat's last theorem, commenced in Chapter 1.

The challenge Fermat left behind, you will recall, is to prove that if n is greater than 2, the equation

$$x^n + y^n = z^n$$

has no (nontrivial) whole-number solutions. Because of our everyday familiarity with the whole numbers, the simplicity of this question might suggest that it should not be too difficult to find a proof. But this impression is illusory. The problem with many such ad hoc questions is that, in order to find an answer, you have to unearth deep and hidden patterns. In the case of Fermat's last theorem, the relevant patterns proved to be many and varied, and very deep indeed. In fact, it is doubtful if more than a few dozen mathematicians in the entire world are in a position to fully understand much of the recent work on the problem. What makes giving a brief sketch of this work worthwhile is that it provides a powerful illustration of the way in which apparently different areas of mathematics turn out to have deep, underlying connections.

The starting point for most of the past fifty years' work on the problem is to recast Fermat's claim as one about rational-number solutions to equations. First, notice that finding whole-number solutions to an equation of the form

$$x^n + y^n = z^n$$

(including the case $n = 2$) is equivalent to finding rational-number solutions to the equation

$$x^n + y^n = 1.$$

For, if you have a whole-number solution to the first equation—say, $x = a$, $y = b$, $z = c$, where a, b, c are whole numbers—then $x = a/c$, $y = b/c$ is a rational-number solution to the second equation. For example,

$$3^2 + 4^2 = 5^2,$$

so $x = 3$, $y = 4$, $z = 5$ is a whole-number solution to the first equation. Dividing through by the solution value for z, namely, 5, you get a rational-number solution to the second equation, namely, $x = \frac{3}{5}$, $y = \frac{4}{5}$:

$$\left(\frac{3}{5}\right)^2 + \left(\frac{4}{5}\right)^2 = 1.$$

And if you have a rational-number solution to the second equation, say, $x = a/c$, $y = b/d$, so that

$$\left(\frac{a}{c}\right)^n + \left(\frac{b}{d}\right)^n = 1,$$

then by multiplying the two solution numbers by the product of their denominators, cd (actually, the least common multiple will do), you get a whole-number solution to the first equation, namely, $x = ad$, $y = bc$, $z = cd$:

$$(ad)^n + (bc)^n = (cd)^n.$$

When Fermat's problem is formulated as one about rational solutions, patterns of geometry and topology may be brought to bear on it. For instance, the equation

$$x^2 + y^2 = 1$$

is the equation for a circle of radius 1, having its center at the origin. Asking for rational-number solutions to this equation is equivalent to asking for points on the circle whose coordinates are both rational. Because the circle is such a special mathematical object—to many, the most perfect of all geometric objects—it turns out to be an easy task to find such points, using the following simple geometric pattern.

Referring to Figure 6.21, start off by choosing some point P on the circle. Any point will do. In Figure 6.21, P is the point $(-1,0)$, since choosing this point makes the problem a bit simpler. The aim is to find points on the circle whose coordinates are both rational. For any point Q on the circle, draw the line from P to Q. This line will cross the y-axis at some point. Let t denote the height of this crossing point above or below the origin. It is then an easy exercise in algebra and geometry to verify that the point Q will have rational coordinates if, and only if, the number t is rational. So, in order to find rational solutions to the original equation, all you need to do is draw lines from the point P to cross the y-axis a rational distance t above or below the origin, and the point Q where your line meets the circle will have rational coordinates. Thus you have a rational solution to the equation.

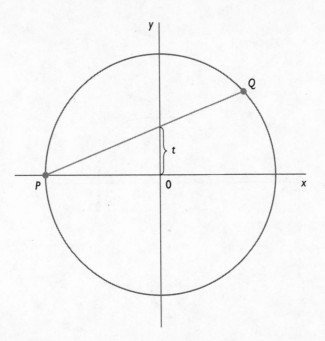

FIGURE **6.21** A geometric method of determining Pythagorean triples.

For instance, if you take $t = \frac{1}{2}$, a little computation shows that the point Q has coordinates $(\frac{3}{5}, \frac{4}{5})$. Similarly, $t = \frac{2}{3}$ leads to the point $(\frac{5}{13}, \frac{12}{13})$, and $t = \frac{1}{6}$ gives $(\frac{35}{37}, \frac{12}{37})$. These correspond to the (whole-number) Pythagorean triples $(3, 4, 5)$, $(5, 12, 13)$, and $(35, 12, 37)$, respectively. In fact, if you analyze this geometric approach, you will see that it leads to the formula for generating all the Pythagorean triples, given on page 45.

In the special case of exponent $n = 2$, therefore, the nice properties of the circle allow you to investigate the rational solutions to the equation

$$x^n + y^n = z^n$$

by means of geometry. But there is no such easy analysis for other values of n, for which the curve is by no means the simple, elegant circle (see Figure 6.22). Recasting the problem in geometric terms, as a search for points on the curve

$$x^n + y^n = 1$$

that have rational coordinates, is still the right direction in which to proceed. But when n is greater than 2, this step is just the start of a long and tortuous path.

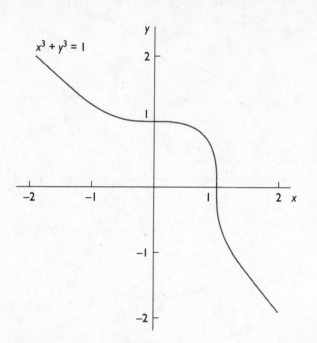

$x^3 + y^3 = 1$

FIGURE **6.22** A Fermat curve: the curve $x^3 + y^3 = 1$.

The problem is that, lacking the nice geometric structure of the circle, the curves you obtain seem no easier to analyze than the original equation. Faced with this hurdle, most mortals would give up and try something else. But if you were one of the many mathematicians whose work has contributed to our present understanding of Fermat's last theorem, you would not. You would press forward, looking for additional structure other than the geometry of the curve. Your hope would be that the increased complexity of this additional structure would provide useful patterns that would help your overall understanding, and eventually lead to a proof.

First of all, you could generalize the problem by allowing *any* polynomial in two unknowns. For any such equation, you could ask whether there are any rational solutions, and maybe by looking at the class of *all* equations, you will be able to discern patterns that will enable you to solve Fermat's original problem. It turns out, however, that this degree of generalization is not sufficient: you need quite a lot more structure before useful patterns start to appear. Curves just do not seem to exhibit enough useful patterns.

So, as a further generalization, suppose that the unknowns x and y in the equation are regarded as ranging not over the real numbers, but over the complex numbers. Then, instead of giving rise to a curve, the equation will determine a surface—a closed, orientable surface, to be precise (see Plate 16 for two examples). Actually, not all equations give rise to a nice, smooth surface, but with some extra effort it is possible to patch things up so that everything works out. The crucial point about this step is that surfaces are intuitive objects that exhibit many useful patterns, and for which there is a wealth of mathematical theory available to be used.

For instance, there is the well-worked-out classification theory for surfaces: every closed, orientable, smooth surface is topologically equivalent to a sphere with a certain number of handles, the number of handles being called the genus of the surface. In the case of a surface arising from an equation, it is natural to call this number the genus of the equation. The genus for the Fermat equation with exponent n works out to be

$$\frac{(n-1)(n-2)}{2}.$$

It turns out that the problem of finding rational solutions to the equation (i.e., rational points on the *curve*) is closely related to the genus of the equation (the genus of the associated *surface*). The larger the genus, the more complicated the geometry of the surface, and the harder it becomes to find rational points on the curve.

The simplest case is that in which the genus is 0, as is true for the Pythagorean-style equations

$$x^2 + y^2 = k,$$

where k is any integer. In this case, one of two results is possible. One possibility is that the equation has no rational points, like the equation

$$x^2 + y^2 = -1.$$

Alternatively, if there is a rational point, then it is possible to establish a one-to-one correspondence between all the rational numbers, t, and all the rational points on the curve, as I did above for the circle. In this case, there are infinitely many rational solutions, and the t-correspondence gives a method for computing those solutions.

The case of curves of genus 1 is much more complicated. Curves determined by an equation of genus 1 are called *elliptic curves*, since they arise in the course of computing the length of part of an ellipse. Examples of elliptic curves are shown in Figure 6.23. Elliptic curves have a number of properties that make them extremely useful in number theory. For example, some of the most powerful known methods for

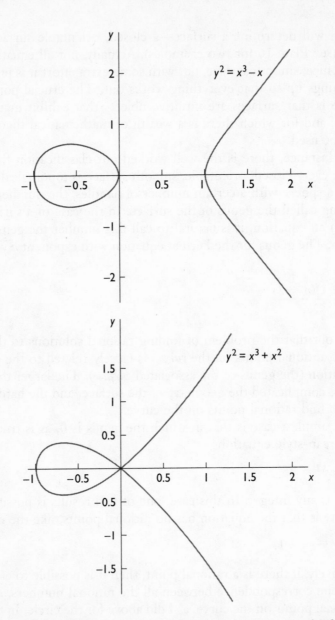

FIGURE **6.23** Two elliptic curves. The curve on the left is the graph of a single function. Even though it breaks apart into two separate pieces, mathematicians refer to it as a single 'curve'. The curve on the right crosses itself at the origin.

factoring large integers into primes (on a computer) are based on the theory of elliptic curves.

As in the case of curves of genus 0, an elliptic curve may have no rational points. But if there is a rational point, then an interesting thing happens, as the English mathematician Lewis Mordell discovered in the early part of the twentieth century. Mordell showed that, although the number of rational points may be either finite or infinite, there is always a *finite* set of rational points—they are called *generators*—such that all other rational points may be generated from them by a simple, explicit process. All that is required is some elementary algebra coupled with the drawing of lines that either are tangent to the curve or cut it at three points. Thus, even in the case where there are infinitely many rational points, there is a structure—a pattern—to handle them.

Of course, the genus 1 case is not in itself of particular interest if the goal is to prove Fermat's last theorem, which, if you discount the case of exponent $n = 3$, concerns equations of genus greater than 1. But, as a result of his investigations, in 1922 Mordell made an observation of tantalizing relevance to Fermat's last theorem: no one had ever found an equation of genus greater than 1 that has infinitely many rational solutions! In particular, all of the many equations that Diophantus had examined turned out to have genus 0 or 1. Mordell proposed that this was not just an accident, but that no equation of genus greater than 1 could have an infinite number of rational solutions.

In particular, Mordell's conjecture implied that, for each value of the exponent n greater than 2, the Fermat equation

$$x^n + y^n = 1$$

can have at most a finite number of rational solutions. Thus, a proof of Mordell's conjecture, while not proving Fermat's last theorem, would mark significant progress toward that goal.

Mordell's conjecture was finally proved in 1983, by a young German mathematician named Gerd Faltings. Faltings had to combine a number of deep ideas in order to produce his proof. The first of these key ideas appeared in 1947 in the work of André Weil, who had been investigating whole-number solutions to equations with respect to finite arithmetics. Weil's basic question was, given a prime number p, how many whole-number solutions does an equation have in modulus p? This question is obviously related to Fermat's last theorem, since if there are no solutions in modulus p, then there can be no solutions in the usual sense. By analogy with some results in topology, Weil formulated a number of technical conjectures concerning this problem. These conjectures involved what are known as *algebraic varieties*—loosely speaking, sets of solutions not to a single equation, but a whole system of equations. Weil's conjectures were finally proved in 1975, by Pierre Deligne.

The second significant contribution to the proof of Mordell's conjecture arose from an analogy between ordinary equations, whose coefficients are numbers, and equations whose coefficients are rational functions—functions of the form $p(x)/q(x)$, where $p(x)$ and $q(x)$ are polynomials. This analogy is a very strong one, and many concepts and results in number theory have analogues in these *function fields*, as they are known. Mordell's conjecture, in particular, has an analogue. When the Soviet mathematician Yuri Manin proved this analogue in 1963, it provided additional evidence that Mordell's conjecture might turn out to be true.

A third ingredient of Faltings's proof was the Shafarevich conjecture. Shortly before Manin obtained his result, his countryman Igor Shafarevich formulated a conjecture concerning the way in which information about the whole-number solutions to an equation can be pieced together from solutions to certain other equations—namely, the equations that result when the original equation is interpreted in mod p finite arithmetic for different primes p. In 1968, A. N. Parshin proved that the Shafarevich conjecture implies Mordell's conjecture.

Meanwhile, in 1966, a fourth contribution was made when the American John Tate came up with yet another conjecture about algebraic varieties. This proliferation of conjectures was a reflection of the growing understanding of the emerging structure of the problem. Generally, mathematicians make a guess public only when they have some intuition to support it. In this case, all the different guesses were heading in the same direction. For his 1983 proof of Mordell's conjecture, Faltings first proved Tate's conjecture. Combining that proof with Deligne's results on Weil's conjectures, he was then able to establish the Shafarevich conjecture. Because of Parshin's 1968 result, this at once yielded Mordell's conjecture, and with it, a proof of the fact that no Fermat equation can have an infinite number of solutions. This process was a marvelous illustration of the way in which increasing abstraction and a search for ever deeper patterns can lead to a proof of a quite concrete result—in this case, a result about whole-number solutions to simple equations.

Three years later, there was another major advance in our understanding of Fermat's last theorem. As with the proof of Mordell's conjecture, an intricate sequence of conjectures was involved, and once again elliptic curves played a significant role in the story.

In 1955, the Japanese mathematician Yutaka Taniyama had proposed a connection between elliptic curves and another well-understood—but not easily described—class of curves, known as modular curves. According to Taniyama, there should be a connection between any given elliptic curve and a modular curve, and this connection should control many of the properties of the elliptic curve.

Taniyama's conjecture was made more precise in 1968 by André Weil, who showed how to determine the exact modular curve that should be connected to a given elliptic curve. In 1971, Goro Shimura demonstrated that Weil's procedure works for a very special class of equations. Taniyama's proposal became known as the Shimura-Taniyama (or sometimes the Shimura-Taniyama-Weil) conjecture.

So far, there was no obvious connection between this very abstract conjecture and Fermat's last theorem, and most mathematicians would have doubted that there was any connection at all. But, in 1986, Gerhard Frey, a mathematician from Saarbrucken, Germany, surprised everyone by finding a highly innovative link between the two.

Frey realized that, if there are whole numbers a, b, c, n such that $c^n = a^n + b^n$, then it is unlikely that one could understand the elliptic curve given by the equation

$$y^2 = x(x - a^n)(x + b^n)$$

in the way proposed by Taniyama. Following an appropriate reformulation of Frey's observation by Jean-Pierre Serre, the American mathematician Kenneth Ribet proved conclusively that the existence of a counterexample to Fermat's last theorem would in fact lead to the existence of an elliptic curve that could not be modular, and hence would contradict the Shimura-Taniyama conjecture. Thus, a proof of the Shimura-Taniyama conjecture would at once imply Fermat's last theorem.

This was tremendous progress. Now there was a definite structure to work with: the Shimura-Taniyama conjecture concerned geometric objects about which a great deal was known—enough to provide good reason to believe the result. There was also a hint of how to set about finding a proof. At least, an English mathematician named Andrew Wiles saw a way to proceed.

Wiles had been fascinated with Fermat's last theorem since he was a child, when he had attempted to solve the problem using high-school mathematics. Later, when as a student at Cambridge University, he learned of Ernst Kummer's work, he tried again, using the German's more sophisticated techniques. But when he realized how many mathematicians had tried and failed to solve the problem, he eventually gave up on it and concentrated on mainstream contemporary number theory, in particular the theory of elliptic curves.

It was a fortuitous choice. For as soon as Ribet proved his astonishing and completely unexpected result, Wiles found himself an acknowledged world expert in the very techniques that could perhaps provide the elusive key to the proof of Fermat's last theorem. For the next seven years he concentrated all his efforts on trying to find a way to prove the Shimura-Taniyama conjecture. By 1991, using powerful new methods developed by Barry Mazur, Matthias Flach, Victor Kolyvagin, and others, he felt sure he could do it.

Two years later, Wiles was sure he had his proof, and, in June 1993, at a small mathematics meeting in Cambridge, England, he announced that he had succeeded. He had, he said, proved the Shimura-Taniyama conjecture, and thus had finally proved Fermat's last theorem. (Strictly speaking, he claimed to have proved a special case of the conjecture that applies not to all elliptic curves, but to a special kind of elliptic curve. The elliptic curves to which his result applied did, however, include those necessary to prove Fermat's last theorem.)

He was wrong. By December of that year, he had to admit that one key step in his argument did not seem to work. Though everyone agreed that his achievement was one of the most significant advances in number theory of the twentieth century, it appeared that he was destined to follow in the footsteps of all those illustrious mathematicians of years gone by, including perhaps Fermat himself, who had dared rise to the challenge laid down in that tantalizing marginal note.

Several months of silence followed, while Wiles retreated to his Princeton home to try to make his argument work. In October 1994 he announced that, with the help of a former student, Richard Taylor of Cambridge University, he had succeeded. His proof—and this time everyone agreed it was correct—was given in two papers: a long one entitled *Modular Elliptic Curves and Fermat's Last Theorem*, which contained the bulk of his argument, and a shorter, second paper co-authored with Taylor, entitled *Ring Theoretic Properties of Certain Hecke Algebras*, which provided a key step he had used in his proof. The two papers together constitute the May 1995 issue of the prestigious research journal *Annals of Mathematics*.

Fermat's last theorem was a theorem at last.

The story of Fermat's last theorem is a marvelous illustration of humanity's never-ending search for knowledge and understanding. But it is much more than that. Mathematics is the only branch of science in which a precise technical problem formulated in the seventeenth century, and having its origins in ancient Greece, remains as pertinent today as it did then. It is unique among the sciences in that a new development does not invalidate the previous theories, but builds on what has gone before. A long path leads from the Pythagorean theorem and Diophantus' *Arithmetic*, to Fermat's marginal comment, and on to the rich and powerful theory we have today, culminating in Wiles's final proof. A great many mathematicians have contributed to that development. They have lived (and are living) all over the world; they have spoken (and speak) many languages; most of them have never met. What has united them has been their love for mathematics. Over the years, each has helped the others, as new generations of mathematicians have adopted and adapted the ideas of their predecessors. Separated by time, space, and culture, they have all contributed to a single enterprise. In this respect, perhaps, mathematics can serve as an example to all humanity.

CHAPTER 7

How Mathematicians Figure the Odds

Every year, thirty million people flock to a small town in the middle of the Nevada desert. The purpose of their trip—and the only reason why the city of Las Vegas is anything other than a small, sleepy, dusty hamlet—is gambling. In present-day America, gambling is a $40 billion business, and it is growing faster than almost any other industry. By understanding the patterns of chance, casinos ensure that they make an average of three cents on every dollar wagered. As a result, they enjoy annual profits of $16 billion.

Contemplating today's glitzy—and sometimes seedy—world of casinos, it is hard to imagine that the entire gambling industry rests upon a series of letters exchanged by two French mathematicians in the middle of the seventeenth century.

Equally dependent on that seventeenth-century mathematics is gambling's more respectable cousin, the insurance industry. (Actually, insurance wasn't always viewed as respectable. Until well into the eighteenth century, the issuance of life insurance was illegal in all European countries except England.)

In mathematical terms, those two Frenchmen founded the present-day subject of *probability theory*—the branch of mathematics that studies the patterns of chance.

Who gets heaven?

People have always been fascinated by chance. According to ancient Greek mythology, the world began when the three brothers Zeus, Poseidon, and Hades rolled dice for the universe. On that occasion, so the story goes, Zeus won the first prize, the heavens, Poseidon took the second prize, the seas, and Hades had to settle for hell.

Early dice generally consisted of the small, squarish knucklebones, called astralagi, taken from the ankles of sheep or deer. Paintings of dice games using astralagi have been found on the walls of Egyptian tombs and on Greek vases, and polished astralagi have been found in archaeological excavations in many parts of the ancient world.

But for all our fascination with dice and other games of chance, it was not until the seventeenth century that anyone worked out the mathematics of these games. Perhaps surprisingly, the Greeks did not even try to develop such a theory. Given their high regard for mathematical knowledge, this was almost certainly because they did not believe there was any order to be found in chance events. To the Greeks, chance was the complete absence of order. Aristotle wrote, "It is evidently equally foolish to accept probable reasoning from a mathematician and to demand from a rhetorician demonstrative proofs."

In some sense, the Greeks were right: there is no order in a purely chance event taken in isolation. In order to find order in chance—to discover a mathematical pattern—you have to see what happens when the same kind of chance event is repeated many times. The order studied in probability theory arises in the repetition of a chance event.

Figuring the odds

The first step toward a theory of chance was taken when a sixteenth-century Italian physician—and keen gambler—named Girolamo Cardano described how to give numerical values to the possible outcomes of a roll of dice. He wrote up his observations in a book titled *Book on Games of Chance*, which he published in 1525.

Suppose you roll a die, said Cardano. Assuming the die is 'honest', there is an equal chance that it will land with any of the numbers 1 to 6 on top. Thus, the chance that each of the numbers 1 to 6 will be face up is 1 in 6, or $\frac{1}{6}$. Today, we use the word *probability* for this numerical value: we say that the probability that the number 5, say, will be thrown is $\frac{1}{6}$.

Going a step further, Cardano reasoned that the probability of throwing either a 1 or a 2 must be $\frac{2}{6}$, or $\frac{1}{3}$, since the desired outcome is one of two possibilities from a total of six.

Going further still—though not quite far enough to make a real scientific breakthrough—Cardano calculated the probabilities of certain outcomes when a die is thrown repeatedly, or when two dice are thrown at once.

For instance, what is the probability of throwing a 6 twice in two successive rolls of a die? Cardano reasoned that it must be $\frac{1}{6}$ times $\frac{1}{6}$, that is, $\frac{1}{36}$. You multiply the two probabilities, since each of the six possible outcomes on the first roll can occur with each of the six possible outcomes on the second roll, yielding thirty-six possible combinations in all. Likewise, the probability of throwing either a 1 or a 2 twice in two successive rolls is $\frac{1}{3}$ times $\frac{1}{3}$, namely, $\frac{1}{9}$.

What is the probability that, when two dice are thrown, the two numbers showing face up will add up to, say, 5? Here is how Cardano analyzed that problem. For each die, there are six possible outcomes. So there are thirty-six (6×6) possible outcomes when the two dice are thrown: each of the six possible outcomes for one of the two dice can occur with each of the six possible outcomes for the other. How many of these outcomes sum to 5? List them all: 1 and 4, 2 and 3, 3 and 2, 4 and 1. That's four possibilities altogether. So of the thirty-six possible outcomes, four give a sum of 5. So, the probability of obtaining a sum of 5 is $\frac{4}{36}$, that is, $\frac{1}{9}$.

Cardano's analysis provided just enough insight for a prudent gambler to be able to bet wisely on the throw of the dice—or perhaps be wise enough not to play at all. But Cardano stopped just short of the key step that leads to the modern theory of probability. So too did the great Italian physicist Galileo, who rediscovered much of Cardano's analysis early in the seventeenth century at the request of his patron, the Grand Duke of Tuscany, who wanted to improve his performance at the gaming tables. What stopped both Cardano and Galileo was that they did not look to see whether there was a way to use their numbers—their probabilities—to predict the future.

That key step was left to the two French mathematicians alluded to at the start of the chapter, Blaise Pascal and Pierre de Fermat. In 1654, the two exchanged a series of letters that most mathematicians today agree was the beginning of the modern theory of probability. Though their analysis was phrased in terms of a specific problem about a gambling game, Pascal and Fermat developed a general theory that could be applied in a wide variety of circumstances to predict the likely outcomes of various courses of events.

The problem that Pascal and Fermat examined in their letters had been around for at least two hundred years: How do two gamblers split the pot if their game is interrupted partway through? For instance, suppose the two gamblers are playing a best-of-five dice game. In the middle of the game, with one player leading two to one, they have to abandon the game. How should they divide the pot?

If the game were tied, there wouldn't be a problem. They could simply split the pot in half. But in the case being examined, the game is not tied. To be fair, they need to divide the pot to reflect the two-to-one advantage that one player has over the other. They somehow have to figure out what would *most likely* have happened had the game been allowed to continue. In other words, they have to be able to look into the future—or in this case, a hypothetical future that never came to pass.

The problem of the unfinished game seems to have first appeared in the fifteenth century, when it was posed by Luca Pacioli, the monk who taught Leonardo de Vinci mathematics. It was brought to Pascal's attention by Chevalier de Méré, a French nobleman with a taste for both gambling and mathematics. Unable to resolve the puzzle on his own, Pascal asked Fermat—widely regarded as the most powerful mathematical intellect of the time—for his help.

To arrive at an answer to Pacioli's puzzle, Pascal and Fermat examined all the possible ways in which the game could have continued, and observed which player won in each case. For instance, suppose Pascal and Fermat are themselves the two players, and that it is Fermat who has the two-to-one lead after game 3. There are four ways the series could be continued to the end: Pascal could win games 4 and 5; or Pascal could win game 4 and Fermat game 5; or Fermat could win games 4 and 5; or Fermat could win game 4 and Pascal game 5. Of course, in practice, in either of the two cases in which Fermat won game 4, the two players would surely stop playing, since then Fermat would have won the series. But mathematically, we have to consider all possible outcomes in the series of *five* games the two set out to play. (This is one of the key insights Fermat and Pascal brought to the solution of the problem.)

Of the four possible ways to complete the entire series, in three of them, Fermat would win the series. (The only way Fermat could lose is if Pascal won both games 4 and 5.) Thus the probability that Fermat would go on to win the series, were they able to continue, is $\frac{3}{4}$. Accordingly, they should split the pot so that $\frac{3}{4}$ goes to Fermat and $\frac{1}{4}$ to Pascal.

Thus, the general approach the two French mathematicians took was to enumerate (and count) all the possible outcomes of the game, and then observe (and count) which of those possible outcomes lead to a particular result (say, a win for one player). As Fermat and Pascal realized, this method could be applied to many other games and sequences of chance events, and in this way, their solution to Pacioli's problem, taken with the work of Cardano, constituted the beginnings of the modern theory of probability.

The geometric patterns of chance

Though Pascal and Fermat established the theory of probability collaboratively, through their correspondence—the two men never met—they each approached the problem in different ways. Fermat preferred the algebraic techniques that he used to such devastating effect in number theory. Pascal, on the other hand, looked for geometric order beneath the patterns of chance. That random events do indeed exhibit geometric patterns is illustrated in dramatic fashion by what is now called *Pascal's triangle*, shown in Figure 7.1.

The symmetrical array of numbers shown in Figure 7.1 is constructed according to the following simple procedure.

- At the top, start with a 1.
- On the line beneath it, put two 1s.
- On the line beneath that, put a 1 at each end, and in the middle put the sum of the two numbers above and to each side, namely, 1 + 1 = 2.
- On line four, put a 1 at each end, and at each point midway between two adjacent numbers on line three put the sum of those numbers. Thus, in the second place you put 1 + 2 = 3 and in the third place you put 2 + 1 = 3.
- On line five, put a 1 at each end, and at each point midway between two adjacent numbers on line four put the sum of those numbers. Thus, in the second place you put 1 + 3 = 4, in the third place you put 3 + 3 = 6, and in the fourth place you put 3 + 1 = 4.

By continuing in this fashion as far as you want, you generate Pascal's triangle. The patterns of numbers in each row occur frequently in probability computations—Pascal's triangle exhibits a geometric pattern in the world of chance.

For example, suppose that when a couple has a child, there is an equal chance of the baby being male or female. (This is actually not quite ac-

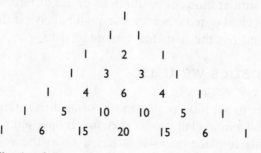

FIGURE **7.1** Pascal's triangle.

curate, but it's close.) What is the probability that a couple with two children will have two boys? A boy and a girl? Two girls? The answers are, respectively, $\frac{1}{4}$, $\frac{1}{2}$, and $\frac{1}{4}$. Here's why. The first child can be male or female, and so can the second. So we have the following four possibilities (in order of birth in each case): boy–boy, boy–girl, girl–boy, girl–girl. Each of these four possibilities is equally likely, so there is a 1 in 4 probability that the couple will have two boys, a 2 in 4 probability of having one child of each gender, and a 1 in 4 probability of having two girls.

Here is where Pascal's triangle comes in. The third line of the triangle reads 1 2 1. The sum of these three numbers is 4. Dividing each number in the row by the sum 4, we get (in order): $\frac{1}{4}$, $\frac{2}{4}$ ($= \frac{1}{2}$), $\frac{1}{4}$, the three probabilities for the different family configurations.

Suppose the couple decides to have three children. What is the probability that they will have three boys? Two boys and a girl? Two girls and a boy? Three girls? The fourth row of Pascal's triangle gives the answers. The row reads 1 3 3 1. These numbers sum to 8. The different probabilities are, respectively: $\frac{1}{8}$, $\frac{3}{8}$, $\frac{3}{8}$, $\frac{1}{8}$.

Similarly, if the couple has four children, the various possible gender combinations have probabilities $\frac{1}{16}$, $\frac{4}{16}$, $\frac{6}{16}$, $\frac{4}{16}$, $\frac{1}{16}$. Simplifying these fractions, the probabilities read: $\frac{1}{16}$, $\frac{1}{4}$, $\frac{3}{8}$, $\frac{1}{4}$, $\frac{1}{16}$.

In general, for any event whose individual outcomes can occur with equal probability, Pascal's triangle gives the probabilities of the different possible combinations that can arise when the event is repeated a fixed number of times. If the event is repeated N times, you look at the $N + 1$ row of the triangle, and the numbers in that row give the numbers of different ways in which each particular combination can arise. Dividing the row numbers by the sum of the numbers in the row then gives the probabilities.

Since Pascal's triangle can be generated by a simple geometric procedure, it shows that there is geometric structure beneath questions of probability. Its discovery was a magnificent achievement. Pascal's triangle can be used to predict the future when the different individual outcomes are all equally likely. But this raises an obvious question: is it possible to make similar kinds of predictions under different circumstances, when the individual outcomes are not equally likely? If the answer is yes, then humankind has the basis for managing risk.

In mathematics we trust

You don't have to set foot in a casino in order to gamble. Indeed, many individuals who would claim to be very much opposed to gambling nevertheless regularly place bets—bets based upon the value they put on their lives, their houses, their cars, and their other possessions. For that is exactly what you are doing when you buy insurance. The insurance company estimates the probability that, say, your car will sustain serious damage in an accident, and then offers you odds against that happening. If there is no accident, the insurer keeps the comparatively small premium you have paid. If there is an accident, the insurer pays out the cost of repairs or a new car.

The system works because the insurer uses mathematics to compute the likelihood of your having an accident. Based on the measured (or estimated) frequency of accidents, the insurer can determine the premiums for the policies sold so that the total amount received in premiums exceeds (by an appropriate margin) the total amount likely to be paid out. If, in one year, there are more claims than anticipated, the insurer will have to pay out more than anticipated, and profits will fall. In a year when there are fewer claims than expected, the company's profits will be higher than normal.

This is a little different from calculating the probabilities of outcomes of gambling games. For games of chance, you can determine the precise probability of each particular outcome, as Cardano did for the throw of dice. In the case of events such as automobile accidents or deaths, however, it is not possible to determine the probabilities by pure reasoning. You have to collect some real data.

For instance, life insurance policies are based on life expectancy tables, which list the number of years a person is likely to live, depending on their current age, where they live, their occupation, their lifestyle, and so forth.

Life expectancy tables are drawn up by making a statistical survey of the population. The first such survey was carried out in London in 1662, by a merchant named John Graunt. He made a detailed analysis of the births and deaths in London between 1604 and 1661, publishing his results in a book called *Natural and Political Observations made upon the Bills of Mortality*. His main source of data was the "Bills of Mortality" that the city of London had started to collect in 1603. The Bills of Mortality recorded all reported deaths in the city, together with their causes, on a week-by-week basis, and listed the number of children christened each week. Graunt paid particular attention to the causes of deaths, among which the plague figured highly, being rampant at the time. It may interest today's city dwellers to observe that in the entire year 1632, which Graunt analyzed, there were only seven murders. That same year, Graunt found one death reported to be caused by being "bit with mad dog" and one resulting from being "affrighted."

Exactly why Graunt decided to carry out his study is not known. It might have been pure intellectual curiosity. He wrote that he found "much pleasure in deducing so many abstruse, and unexpected infer-ences out of these poor despised Bills of Mortality." On the other hand, he also seems to have had a business objective. He wrote that his re-search enabled him "to know how many people there be of each Sex, State, Age, Religious, Trade, Rank, or Degree, &c. by the knowing whereof Trade and Government may be made more certain, and Regu-lar; for, if men know the People as aforesaid, they might know the con-sumption they would make, so as Trade might not be hoped for where it is impossible." Whatever his motivation, Graunt's work was one of the very first examples of modern statistical sampling and market research.

Thirty years after Graunt published his findings, the English as-tronomer Edmund Halley—most famous for his discovery of the comet that bears his name—carried out a similar, though much more intensive, analysis of mortality figures. Halley's data, which were very detailed, came from the town of Breslau, Germany—the present-day Wrozlaw in Poland—where they had been collected on a monthly basis between 1687 and 1691.

With the methods of data collection and analysis developed by Graunt and Halley, the stage was set for the development of the mod-ern insurance business. Starting with mortality data, for example, it was possible to calculate, for each category of person (categorized by age, gender, wealth, occupation, etc.), the probability that an individual would die during the coming year. This meant that mathematicians could use the theory of probability to predict future events, such as the death of a particular person in the coming year. Because of the depen-dence on collected data, the use of probability theory in such circum-stances would not be as reliable as when applied to the gaming tables, where the probability of, say, the outcome of a particular roll of dice can be computed precisely. Nevertheless, predictions of future events could be made with sufficient reliability to serve as the basis of a profitable in-surance industry. However, largely due to moral objections, it was not until the eighteenth century that insurance companies started to appear.

The first American insurance company was called, appropriately, First American. It was a fire insurance company, established by Ben-jamin Franklin in 1752. The first life insurance policies in America were issued by the Presbyterian Ministers' Fund in 1759. The word 'policy', in-cidentally, comes from the Italian word 'polizza', which means a promise.

One of the first international insurance companies, still active today, was the famous Lloyd's of London, which was started in 1771 when sev-enty-nine individual insurers entered into a collaborative agreement. The name they chose for their new company was that of the establish-ment where they had hitherto carried out their business—mostly ship-ping insurance: Edward Lloyd's coffee house on Lombard Street in London. Lloyd himself played an active part in this development: in 1696, five years after he had opened the coffee house, he started "Lloyd's List," a compilation of up-to-date information on the arrival and departure of ships and conditions at sea and abroad—information of obvious importance to anyone seeking to insure ships or cargo.

Today, insurance companies offer policies to cover all kinds of eventualities: death, injury, auto accident, theft, fire, flood, earthquake, hurricane, accidental damage to household goods, loss of luggage on an airplane, et cetera. Movie actors insure their looks, dancers insure their legs, and singers insure their voices. You can even buy insurance to cover rain on the day of a barbecue or something going wrong at your wedding.

But we are getting ahead of our story. Considerable mathematical advances were made between the statistical work of Graunt and the establishment of the present-day insurance industry.

The Amazing Bernoullis

The development of the mathematics of chance during the eighteenth century owes much to two members of one of the most remarkable families of all time. "The Amazing Bernoullis" sounds like the name of a circus act. However, it was not on the trapeze or the high wire that the Bernoullis performed their dazzling feats, but in mathematics.

In all, there were eight members of the family who distinguished themselves in mathematics. The father of the family was Nicolaus Bernoulli, a wealthy merchant who lived in Basel, Switzerland, from 1623 to 1708. His three sons Jacob, Nicolaus, and Johann all went on to become first-rate mathematicians.

One of Jacob's most significant contributions to mathematics is his 'law of large numbers', a major theorem in the theory of probability. The other members of the family who contributed to the new mathematics of chance were Johann's son Daniel, also known for his discovery of 'Bernoulli's equation', which keeps an aircraft in the sky, and Jacob's nephew Nicolaus (another Nicolaus—there were four Nicolauses in all). Jacob and Daniel were interested in different aspects of what was essentially the same question: how can probability theory be taken from the gaming table—where exact probabilities can be calculated—and applied in the far more messy real world?

The specific problem that Jacob worked on was already implicit in the earlier work of John Graunt on mortality rates. Graunt was well aware that the data he had available, while extensive, represented only a sample of the entire population: the population of London. And even for the population of London, the data were for a specific time period. Yet the relatively limited nature of the data did not prevent him from making generalizations that went beyond the data itself. By extrapolating from the data in the Bills of Mortality to obtain conclusions for the entire country and for more extensive periods of time, Graunt was one of the first analysts to perform what is nowadays called *statistical inference*: drawing conclusions about a large population based on data collected for a small sample. For such a process to lead to reliable conclusions, the sample should be 'representative' of the whole population. How can a representative sample be chosen?

Another related question is, will a larger sample ensure a more reliable result, and if so, how large does the sample have to be? Jacob Bernoulli raised this very question in a letter to his friend Leibniz (he of the calculus) in 1703.

In his pessimistic reply, Leibniz observed that "nature has established patterns originating in the return of events, but only for the most part." It was the phrase "only for the most part" that seemed to stand in the way of a mathematical analysis of many 'real-life' probabilities.

Undeterred by the discouraging tone of Leibniz's reply, Bernoulli continued his investigation. In what were to be the last two remaining years of his life, he made considerable progress. After he died in 1705, his nephew Nicolaus Bernoulli organized his uncle's results into publishable form, though so challenging was the task that it took him eight years before the book, *The Art of Conjecture*, was published in 1713, with Jacob given as the author.

Jacob began by recognizing that the question he had asked Leibniz really split into two parts, depending on two different notions of probability. First, there was what Jacob called *a priori probability*—probability computed before the fact. Was it possible to compute a precise probability of the outcome of an event before the event occurred? In the case of games of chance, the answer was yes. But as Leibniz rightly pointed out, probabilities computed in advance for events such as illness or death could be reliable "only for the most part."

Jacob used the term *a posteriori probability* to refer to the other kind of probability—probability computed after the event. Given a sample of a population, if a probability is computed for that sample, how reliably does that probability represent that for the entire population? For example, suppose you are given a large (opaque) jar of red and blue marbles. You know there are 5,000 marbles in all, but you don't know how many there are of each color. You draw a marble from the jar at random, and it is red. You put it back, shake up the jar, and then draw again. This time you draw a blue marble. You repeat the process—shaking, drawing, and returning—50 times, and find that 31 times you draw a red marble and 19 times a blue marble. This leads you to suspect that there are around 3,000 red marbles and 2,000 blue ones, so that the a posteriori probability of drawing a red marble at random is $\frac{3}{5}$. But how confident can you be in this conclusion? Would you be more confident if you had sampled a greater number of marbles—say, 100?

Bernoulli proved that by taking a sufficiently large sample, you can increase your confidence in the computed probability to whatever degree you wish. More precisely, by increasing the sample size, you can increase to any level your confidence that the probability computed for the sample is *within any stipulated amount* of the true probability. This is the result called the *law of large numbers*, and it is a central theorem of probability theory.

In the case of the 5,000 marbles in the jar, if there are exactly 3,000 red ones and exactly 2,000 blue, Bernoulli calculated how many marbles you would have to sample to be certain to within one in a thousand that the distribution you found in the sample is within 2 percent of the true ratio of 3:2. The answer he obtained was 25,550 drawings. That's far more than the total number (5,000) of marbles in the first place. So, in this example, it would be much more efficient simply to count all the marbles! Nevertheless, Bernoulli's theoretical result does show that by taking a sufficiently large sample, probabilities can be computed from a sample so that to any degree of certainty (other than absolute certainty), the computed probability is within any desired degree of accuracy of the true probability.

Even if you have computed an a posteriori probability, how reliable a guide is that probability for predicting future events? This is not really a mathematical question. The issue is, rather, can the past be taken as indicative of the future? The answer can vary. As Leibniz remarked in his pessimistic letter to Jacob Bernoulli, "New illnesses flood the human race, so that no matter how many experiments you have done on corpses, you have not thereby imposed a limit on the nature of events so that in future they could not vary." More recently, the late Fischer Black, a pioneering financial expert who exchanged his academic position at M.I.T. for the financial market on Wall Street, made a similar remark about predicting the future on the basis of a mathematical analysis of the past: "Markets look a lot less efficient from the banks of the Hudson than the banks of the Charles."

Fear of flying

As a friend of Leibniz, Jacob Bernoulli was one of the first mathematicians to learn about the remarkable new methods of the calculus—the study of motion and change. Indeed, much of the early development of the subject was carried out by Bernoulli.

Daniel Bernoulli too did pioneering work on the calculus. He took a different tack from his uncle Jacob by applying the methods of the calculus to flowing liquids and gases. Arguably the most significant of his many discoveries was what is nowadays known as 'Bernoulli's equation', the equation that keeps an aircraft aloft, and which forms the basis of all modern aircraft design.

Though the application of Daniel's equation to the airline industry of the twentieth century was two hundred years away, other work of the great Swiss mathematician would also turn out to be relevant to the world of air travel. Daniel's contributions to our understanding of probability are highly relevant to that familiar and oft-cited fact that, although modern air travel is the safest way to travel, many people nevertheless feel extremely nervous when they board a plane. Indeed, for some people, the fear of flying is so great that they never fly at all.

Such individuals are not necessarily unaware of the odds. They may well know that the probability of being involved in an accident is extremely small, far less than when they travel by car, for example. Rather, the issue is the nature of an airline crash and the importance they attach to such an event, however unlikely it may be.

Fear of lightning is a similar phenomenon, in which the tiny mathematical probability of being struck by a lightning bolt is far outweighed by the significance many individuals attach to such an event.

It was this essentially human aspect of probability theory that interested Daniel Bernoulli: was it possible to make specific observations about the way in which people actually assessed risk? In 1738, he published a seminal paper on the issue in the *Papers of the Imperial Academy of Sciences in Saint Petersburg*. In that paper he introduced the key new concept of utility.

To appreciate Bernoulli's concept of utility, you need to know something about another idea of probability theory: expectation. Suppose I challenge you to throw a die. If you throw an even number, then I will pay you a dollar amount equal to the number you throw. If you throw an odd number, you will pay me $2. Your *expectation* for the game is a measure of what you can 'expect' to win. It is the average amount you would win per game if you were to play repeatedly.

To compute your expectation, you take the probability of each possible outcome and multiply it by the amount you would win in that case, and then add together all those amounts. Thus, since a loss of $2 can be regarded as a 'win' of –$2, your expectation for the game is:

$$\frac{1}{6} \times \$2 + \frac{1}{6} \times \$4 + \frac{1}{6} \times \$6 + \frac{1}{2} \times (-\$2) = \$1.$$

The individual terms in the sum correspond to your winning $2, $4, and $6 and losing $2, respectively.

The above calculation tells you that, if you were to play the game repeatedly, your average winnings would be $1 per game. Obviously, this game is to your advantage. If I were to change the rules by demanding $4 from you each time you throw an odd number, then your expectation would fall to zero, and the game would be to nobody's advantage. If the penalty you paid for throwing an odd number were more than $4, then, in the long run if not the short, you would most likely lose money.

By taking into account both the probabilities and the payoffs, the expectation measures the value to an individual of a particular risk or wager. The greater the expectation, the more attractive the risk.

At least, that was the theory. And for many examples, expectation seemed to work well enough. But there was a problem, and it was most dramatically illustrated by a tantalizing puzzle proposed by Daniel's cousin Nicolaus. It came to be known as the Saint Petersburg paradox. Here it is.

Suppose I challenge you to a game of repeated coin tosses. If you throw a head on the first toss, I pay you $2, and the game is over. If you throw a tail on the first toss and a head on the second, I pay you $4, and the game is over. If you throw two tails and then a head, I pay you $8, and the game is over. We continue in this fashion until you throw a head. Each time you throw a tail and the game continues, I double the amount you will win if and when you throw a head.

Now imagine that someone else comes along and offers to pay you $10 to take your place in the game. Would you accept or decline? What if he offered you $50? Or $100? In other words, how much do you judge the game to be worth to you?

This is exactly what the expectation is supposed to measure. So what does it work out to be in this case? Well, in principle, the game can go on indefinitely—there are infinitely many possible outcomes: H, TH, TTH, TTTH, TTTTH, . . . The respective probabilities of these outcomes are $\frac{1}{2}, \frac{1}{4}, \frac{1}{8}, \frac{1}{16}, \frac{1}{32}, \ldots$ Thus, the expectation is

$$\frac{1}{2} \times 2 + \frac{1}{4} \times 4 + \frac{1}{8} \times 8 + \frac{1}{16} \times 16 + \frac{1}{32} \times 32 + \ldots$$

This infinite sum can be rewritten as

$$1 + 1 + 1 + 1 + 1 + \ldots$$

Since the sum goes on forever, the expectation is therefore infinite.

According to the theory, faced with an infinite expectation, you should not give up your opportunity to play this game for any amount of money. But most people—even knowledgeable probability theorists—would be tempted to take the $10 offer, and would almost certainly take an offer of $50. The chance of winning as much in the game seems too remote. So what is wrong with the notion of expectation in this case?

Pondering this question, as well as a number of other problems he saw with expectation, led Daniel Bernoulli to replace this highly mathematical concept with the far less formal idea of *utility*.

Utility is intended to measure the significance you attach to a particular event. As such, utility is very much an individual thing. It depends on the value a person puts on a particular outcome. Your utility and mine might differ.

At first glance, replacing the mathematically precise concept of expectation with the decidedly personal idea of utility might appear to render impossible any further scientific analysis. But such is not the case. Even for a single individual, it may well be impossible to assign specific numerical values to utility. Nevertheless, Bernoulli was able to make a meaningful—indeed, profound—observation about utility. He wrote: "[The] utility resulting from any small increase in wealth will be inversely proportionate to the quantity of goods previously possessed."

Bernoulli's utility law explains why even a moderately wealthy person will generally find the pain of losing half his fortune much greater than the pleasure of doubling it. As a result, few of us are prepared to gamble half our wealth for the chance of doubling it. Only when we are truly able to declare "What have I got to lose?" are most of us prepared to take a big gamble.

For instance, suppose you and I each have a net worth of $10,000. I offer you a single toss of a coin. Heads and I give you $5,000; tails and you give me $5,000. The winner comes out with $15,000, the loser with $5,000. Since the payoffs are equal and the probability of each of us winning is $\frac{1}{2}$, we each have an expectation of zero. In other words, according to expectation theory, it makes no difference to either of us whether we play or not. But few of us would play this game. We would almost certainly view it as 'taking an unacceptable risk'. The 0.5 probability of losing $5,000 (half our wealth) far outweighs the 0.5 probability of winning $5,000 (an increase of 50 percent).

The utility law likewise resolves the Saint Petersburg paradox. The longer the game goes on, the greater the amount you will win when a head is finally tossed. (If the game goes on for six tosses, you are bound to win over $100. Nine tosses and your winnings will exceed $1,000. If it lasts for fifty tosses, you will win at least a million billion dollars.) According to Bernoulli's utility law, once you reach the stage at which your minimum winning represents a measurable gain *in your terms*, the benefit to be gained by playing longer starts to decrease. That determines the amount for which you would be prepared to sell your place in the game.

So much for expectation, according to which you should stay in the game as long as you can. In fact, a similar fate was to befall Bernoulli's utility concept in due course, when mathematicians and economists of a later generation looked more closely at human behavior. But the fact remains that it was Daniel Bernoulli who first insisted that if you wanted to apply the mathematics of probability theory to real-world problems, you had to take the human factor into account. On their own, the patterns of chance that can be observed by looking at rolls of dice or tosses of a coin are not enough.

Ringing in the bell curve

With his law of large numbers, Jacob Bernoulli had shown how to determine the number of observations needed in order to ensure that the probability determined from a sample of a population was within a specified amount of the true probability. This result was of theoretical interest, but not a great deal of use in practical applications. For one thing, it required that you know the true probability in advance. Second, as Bernoulli himself had shown in the case of the marbles in the jar, the number of observations required to achieve a reasonably accurate answer could be quite large. Potentially of much greater use was a solution to the opposite problem: given a specified number of observations, can you calculate the probability that they will fall within a specified bound of the true value? A solution to this problem would allow a probability for a population to be calculated with known accuracy on the basis of a probability computed for a sample.

The first person to investigate this question was Jacob's nephew Nicolaus, who worked on it while in the process of completing his late uncle's work and preparing it for publication. To illustrate the problem, Nicolaus gave an example concerning births. Assuming a ratio of 18 male births to every 17 female births, for a total of 14,000 births, the expected number of male births would be 7,200. He calculated that the odds were over 43 to 1 that the actual number of male births in the population would lie between 7,200 − 163 and 7,200 + 163, i.e., between 7,037 and 7,363.

Nicolaus did not completely solve the problem, but made sufficient progress to publish his findings in 1713, the same year his deceased uncle's book finally appeared. Some years later, his ideas were taken up by Abraham de Moivre, a French mathematician who, as a Protestant, had fled to England in 1688 to escape persecution by the Catholics. Unable to secure a proper academic position in his adopted country, de Moivre made a living by tutoring in mathematics and consulting for insurance brokers on matters of probability.

De Moivre obtained a complete solution to the problem Nicolaus Bernoulli had addressed, which he published in 1733 in the second edition of his book *The Doctrine of Chances*. Using methods of the calculus as well as probability theory, de Moivre showed that a collection of random observations tended to distribute themselves around their average value.

These days, de Moivre's distribution is known as the *normal distribution*. When the normal distribution is represented graphically, plotting the observations along the horizontal axis and the frequency (or probability) of each observation along the vertical axis, the resulting curve is shaped like a bell (see Figure 7.2). For this reason, it is often called the *bell curve*.

As the bell curve shows, the largest number of observations tend to be clustered in the center, around the average for all the observations. Moving out from the middle, the curve slopes down symmetrically, with an equal number of observations on either side. The curve slopes slowly at first, then much more rapidly, finally becoming very flat toward the two edges. In particular, observations far from the average are much less frequent than observations close to the average.

De Moivre discovered the bell curve by examining the behavior of random observations. Its elegant symmetrical shape showed that beneath randomness lay an elegant geometry. From a mathematical point of view, that alone was a significant result. But that was not the end of the story. Eighty years later, Gauss noticed that when he plotted large

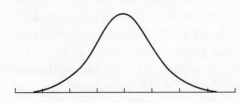

FIGURE **7.2** The bell curve.

numbers of geographic and astronomical measurements, the resulting curves always looked remarkably like de Moivre's bell curve. Different measurements of a particular distance on the earth's surface, for example, or of an astronomical distance, would form a bell-shaped cluster around a central value—the mean (i.e., average) of the individual measurements. The inevitable errors in taking measurements gave rise, it seemed, to a normal distribution of values.

By considering the bell curve not as a geometric feature of randomness, but rather as an inevitable consequence of errors in measurement, Gauss realized that he could use the normal distribution to assess the value of data. In particular, using the bell curve, it would be possible to assign (approximate) numerical probabilities to observations, analogously to the way in which they could be assigned to outcomes of a roll of dice. The closer a particular data value was to the mean, as measured on the bell curve, the greater the probability of that value being correct. The techniques of probability theory developed by Pascal, Fermat, the Bernoullis, and others could now be transferred from the gaming table to other areas of life. As a result of Gauss's contribution, the normal distribution is also sometimes referred to as the Gaussian distribution.

The key technical concept that Gauss required to use the bell curve in this manner was a measure that de Moivre himself had introduced. Known nowadays as the *standard deviation*, de Moivre's measure allows us to judge whether a set of observations is sufficiently representative of the entire population. The standard deviation measures the dispersion of the observations around the mean. For a normal distribution, approximately 68 percent of the observations fall within one standard deviation of the mean, and 95 percent fall within two standard deviations of the mean. For this reason, when newspapers and magazines publish results of surveys, they generally include a statement of the deviation. If they don't, you should view the quoted results with suspicion. Remember, a man whose head is in the oven and whose feet are in the refrigerator could be said to feel just fine on average. But what a deviation in data points!

With statistics playing such a major role in twentieth-century life, the bell curve is an icon of the present age. It allows us to assign numerical probabilities to events, and thereby to apply the methods of probability theory to many areas of life. In general, whenever there is a large population that produces data, with the data produced by each member of the population independent of that produced by the others, the bell curve will arise, and Gauss's method may be used to provide a measure of the reliability of that data.

Following Gauss's lead, insurance companies now use the bell curve to set their rates, business leaders use it to plan expansions into new markets, government officials use it to determine public policy, educators use it to grade student examinations, opinion pollsters use it to analyze their data and make their predictions, medical researchers use it to test the effectiveness of different treatments, economists use it to analyze economic performance, biologists use it to study the growth of plants, and psychologists use it to analyze the behavior of their fellow human beings.

The Reverend Bayes

In today's data-rich society, statistical (or probabilistic) inference plays a significant role in our everyday lives. Much of it remains hidden from our view and beyond our control, and we are unable to respond to it. But what about the statistics we are aware of? How well are we able to make sense of the masses of quantitative and statistical information that bombard us every day through newspapers, magazines, radio, and television and in our workplaces? How well do we fare when we find ourselves having to evaluate statistical information to make important decisions about our health, our homes, or our jobs?

The answer is that when it comes to making sense of statistical data, we often perform very poorly. Hundreds of thousands of years of evolution have equipped us with many useful mental abilities—our instinct to avoid many dangerous situations and our use of language are two obvious examples. However, evolution has not equipped us to handle statistical or probabilistic data—a very recent component of our lives. Where quantitative data is concerned, if we want to reach wise decisions, we often have to rely on mathematics. Thanks to Gauss's work on the bell curve, we can turn to the methods of probability theory. When we do so, we sometimes find that our intuitions are wildly misleading.

For example, suppose that you undergo a medical test for a relatively rare cancer. The cancer has an incidence of 1 percent among the general population. Extensive trials have shown that the reliability of the test is 79 percent. More precisely, although the test does not fail to detect the cancer when it is present, it gives a positive result in 21 percent of the cases in which no cancer is present—what is known as a 'false positive'. When you are tested, the test produces a positive result. What is the probability that you actually have the cancer?

If you are like most people, you will assume that if the test has a reliability rate of nearly 80 percent, and you test positive, then the likelihood that you do indeed have the cancer is about 80 percent (i.e., the probability is approximately 0.8). Are you right?

The answer is no. Given the scenario just described, the likelihood that you have the cancer is a mere 4.6 percent (i.e., the probability is 0.046). Still a worrying possibility, of course. But hardly the scary 80 percent you thought at first.

How do you arrive at such a figure? The mathematics you need was developed by an eighteenth-century English minister named Thomas Bayes.

Specifically, Bayes's method shows you how to calculate the probability of a certain event E (in the above example, having the cancer) based on evidence (in our case, the result of the medical test) when you know:

1. the probability of E in the absence of any evidence
2. the evidence for E
3. the reliability of the evidence (i.e., the probability that the evidence is correct)

In our cancer example, the probability (1) is 0.01, the evidence (2) is that the test came out positive, and the probability (3) has to be computed from the 79 percent figure given. All three pieces of information are highly relevant, and to evaluate the probability that you have the cancer, you have to combine them in the right manner. Bayes's method tells you how to do this.

Before we go through the details, however, here is another scenario in which our intuitions are likely to mislead us, but where Bayes's mathematics once again comes to our rescue.

A certain town has two taxi companies, Blue Cabs and Black Cabs. Blue Cabs has 15 taxis, Black Cabs has 85. Late one night, there is a hit-and-run accident involving a taxi. All of the town's 100 taxis were on the streets at the time of the accident. A witness sees the accident and claims that a blue taxi was involved. At the request of the police, the witness undergoes a vision test under conditions similar to those on the night in question. Presented repeatedly with a blue taxi and a black taxi, in random order, he shows he can successfully identify the color of the taxi four times out of five. (The remaining one-fifth of the time, he misidentifies a blue taxi as black or a black taxi as blue.) If you were investigating the case, which taxi company would you think is most likely to have been involved in the accident?

Faced with evidence from an eyewitness who has demonstrated that he is right four times out of five, you might be inclined to think it was a blue taxi that the witness saw. You might even think that the odds in favor of its being a blue taxi were exactly 4 out of 5 (i.e., a probability of 0.8), those being the odds in favor of the witness being correct on any one occasion.

Bayes's method shows that the facts are quite different. Based on the data supplied, the probability that the accident was caused by a blue taxi is only 0.41. That's right—less than half. The accident is more likely to have been caused by a black taxi.

What human intuition often ignores, but Bayes's method takes proper account of, is the 0.85 probability (85 out of a total of 100) that *any* taxi in the town is likely to be black.

Without the testimony of the witness, the probability that a black taxi caused the accident would have been 0.85, the proportion of taxis in the town that are black. So, *before* the witness testifies to its color, the probability that the taxi in question was blue is low, namely, 0.15. This figure (item 1 in our list above) is what is called the *prior probability* or the *base rate*: the probability based purely on the way things are, not on any particular evidence pertaining to the case in question.

When the witness testifies to its color, that evidence increases the probability that the taxi in question was blue from the 0.15 prior probability figure, but not all the way to the 0.8 figure of the witness's tested accuracy. Rather, the reliability of the witness's evidence (0.8) must be *combined* with the prior probability (0.15) to give the real probability. Bayes's method gives us an exact mathematical way of doing this. It tells us that the correct probability is given by the following sum, where $P(E)$ denotes the probability of event E occurring:

$$\frac{P(\text{blue taxi}) \times P(\text{witness is right})}{P(\text{blue taxi}) \times P(\text{witness is right}) + P(\text{black taxi}) \times P(\text{witness is wrong})}.$$

Putting in the various figures, this becomes

$$\frac{0.15 \times 0.8}{0.15 \times 0.8 + 0.85 \times 0.2},$$

which works out to be

$$\frac{0.12}{[0.12 + 0.17]}$$
$$= \frac{0.12}{0.29}$$
$$= 0.41.$$

How exactly is the above formula derived?

The witness claims that the taxi he saw was blue. He is right $\frac{8}{10}$ of the time. Hypothetically, if he were to try to identify each taxi in turn under the same circumstances, how many would he identify as being blue?

Of the 15 blue taxis, he would (correctly) identify 80 percent as being blue, namely, 12. (In this hypothetical argument, we are assuming that the actual numbers of taxis accurately reflect the probabilities. This is a reasonable assumption for such an argument.)

Of the 85 black taxis, he would (incorrectly) identify 20 percent as being blue, namely, 17. (Again, we assume that the probabilities give the actual numbers.)

So, in all, he would identify 29 of the taxis as being blue. Thus, on the basis of the witness's evidence, we find ourselves looking at a group of 29 taxis.

Of the 29 taxis we are looking at, 12 are in fact blue.

Consequently, the probability of the taxi in question being blue, given the witness's testimony, is $\frac{12}{29}$, or 0.41.

Now let's return to the example of the test for the rare cancer. Recall that the cancer has an incidence of 1 percent in the population, and that the test has a reliability of 79 percent—more precisely, it detects the cancer if it is present, but in 21 percent of cases in which there is no cancer present, it gives a false positive result.

Let's argue as we did for the taxicab example. To keep the arithmetic simple, we'll assume a total population of 10,000 people. Since all we are ultimately concerned about is percentages, this simplification will not affect the final answer. As with the taxicab example, we'll also assume that the various probabilities are reflected exactly in the actual numbers. Thus, of the total population of 10,000, 100 will have the cancer, 9,900 will not.

In the absence of the test, all you could say about the likelihood of your having the cancer is that there is a 1 percent chance that you do.

Then you take the test, and it comes back positive. How do you revise the probability that you have the cancer?

First, there are 100 individuals in the population who do have the cancer, and for all of them, the test will correctly give a positive result, thereby identifying those 100 individuals as having the cancer.

Turning to the 9,900 cancer-free individuals, for 21 percent of them, the test will give a false positive result, thereby incorrectly identifying 9,900 × 0.21 = 2,079 individuals as having the cancer.

Thus, in all, the test identifies a total of 100 + 2,079 = 2,179 individuals as having the cancer. Having tested positive, you are among that group. (This is precisely what the evidence of the test tells you.) The question is, are you in the subgroup that really does have the cancer, or is your test result a false positive?

Of the 2,179 people identified by the test, 100 really do have the cancer. Thus, the probability of your being among that group is $100/2,179 = 0.046$.

In other words, there is a 4.6 percent possibility that you actually have the cancer. Those are still worrisome odds, of course. But they're not at all as bad as the scary 79 percent you thought at first. This computation should indicate why it is important to take the prior probability—in this case, the overall incidence of the cancer in the population—into account.

Enter the average man

The dawn of the age of statistics also saw the birth of a new creature: 'the average man'. You will never, of course, meet this individual in the street—he does not live and breathe and move among us. Like the equally mythical 'average American family' with its 2.4 children that followed him, the average man is a product of the statistician, a child of the bell curve. He first appeared in 1835, in a book called *A Treatise on Man and the Development of His Faculties*, written by a Belgian polymath named Lambert Quetelet.

Quetelet was an early enthusiast of the new science of statistics. He wrote three books on the subject, and helped found several statistical societies, including England's Royal Statistical Society and the International Statistical Congress. '*L'homme moyen*' was Quetelet's name for the numerical creation that provided the general public with a personification of the statistician's analysis of society.

Whereas de Moivre had discovered the bell curve by analyzing random data, and Gauss had shown how it applies to astronomical and geographic measurements, Quetelet brought it into the human and social realm. He found his average man at the midpoint of the bell curve. By collecting massive amounts of data, he listed the different physical, mental, and behavioral attributes of the average man among various groups—selected by age, by occupation, by ethnic origin, and so forth.

Quetelet's efforts were fueled by two motives. First, he had an insatiable appetite for statistics. He could never resist an opportunity to count and to measure. He examined mortality by age, by profession, by locality, by season, in prisons, and in hospitals. He collected statistics on drunkenness, insanity, suicides, and crime. On one occasion he took the chest measurements of 5,738 Scottish soldiers. On another, he tabulated the heights of 100,000 French military conscripts.

Second, he hoped to influence social policy. He wanted to use statistics to identify the causes that led to people being in one grouping and not another. To this end, whenever he collected new data, he would set about organizing it so as to exhibit the bell curve he was sure would be revealed.

In fact, Quetelet was so convinced of the universality of the bell curve that occasionally he produced a normal distribution where none was really present. As present-day statisticians know full well, by manipulating the data enough, it is generally possible to produce any pattern you want. But objectivity is the first casualty of such a process. Collecting data in order to support a cause already committed to, or a course of action already decided upon, is fraught with danger.

His methodological lapses aside, however, Quetelet was the first person to apply the normal distribution to social data, and to attempt to use mathematics to determine the causes of social factors. As such, he was a pioneer of the statistically informed public policy decision making that is a feature of twentieth-century society.

Hot on Quetelet's heels came an Englishman with an equal passion for statistics: John Galton. A cousin of Charles Darwin, Galton established what he called the Anthropometric Laboratory in 1884, with the goal of measuring the human body in every detail. It was largely the laboratory's work on fingerprints—in particular, the book on the subject that Galton wrote in 1893—that led to the widespread use of fingerprinting by the police.

Like Quetelet before him, Galton was impressed by the elegant symmetry of the bell curve, which, like Quetelet, he saw as being ubiquitous. In one experiment, he plotted 7,634 final examination grades in mathematics at Cambridge, obtaining a perfect bell curve. He found another normal distribution in the exam results of the applicants for admission to the Royal Military College at Sandhurst.

Also like Quetelet, Galton pursued his statistics with a specific—though different—goal in mind. Galton was a member of the talented Darwin family. His grandfather was Erasmus Darwin, a highly respected physician and polymath whose book *Zoonomia, or the Theory of Generations*, published in 1796, contains many of the ideas made famous sixty-three years later by his more famous grandson Charles in *The Origin of Species*. Growing up surrounded by what he saw as the intellectual *crème de la crème*, and being exposed to his grandfather's ideas on heredity, Galton believed that special talents were inherited. He set out to make a study of inherited traits, publishing his first set of findings in 1869 in a book titled *Hereditary Genius*. In 1883 he coined the name 'eugenics' for the line of study he was pursuing, a term that was later to take on an unpleasant meaning when used as a 'scientific' justification for racial policies such as those pursued by the Nazis.

In fact, the use of eugenics in order to promote the development of 'master races' flew in the face of the most significant result Galton himself discovered. He called it 'reversion'. Today it is known as *regression to the mean.*

Regression to the mean is the statistical equivalent of the colloquialism "What goes up, must come down." It is the tendency of the members of any population, over time, to gravitate toward the middle of the bell curve—toward the statistical mean.

For instance, while in one study Galton found that a high proportion of close male relatives of 286 judges were themselves judges, admirals, generals, poets, novelists, and physicians, in other studies he observed that such groupings of talent do not persevere through future generations. In one such study of eminent men, for example, he discovered that only 36 percent of their sons and 9 percent of their grandsons achieved eminence. The bell curve, it seemed, represented more of a trap than a prescription for eugenics.

In another study, Galton weighed and measured thousands of sweet peas to find ten specimens each of seven different diameters. He separated the specimens into ten sets, each containing one pea of each size. Then, keeping one specimen set himself, he sent the other nine to friends spread throughout the British Isles, with instructions to plant them under carefully specified conditions. When the plants had grown to maturity, the friends were to return the offspring peas to Galton so that he could compare their sizes with the sizes of their respective parents.

Though he did find evidence that larger parents tended to produce larger offspring, the most significant feature of Galton's results was that they confirmed regression to the mean. The peas in Galton's original sample ranged in size from 15 hundredths of an inch to 21 hundredths of an inch. The peas he received back had a much smaller range, from 15.4 hundredths of an inch to 17.3 hundredths of an inch.

Galton obtained similar results in a subsequent study he made of the 928 adult children of 205 pairs of parents.

There remained little doubt: regression to the mean was a fact. But why did it arise? Though he did not know the mechanism, Galton suspected that random effects were the cause. To test this suspicion, he designed an experimental device known nowadays as a Galton board.

A Galton board consists of a thin sandwich of wood and glass into which small balls can be dropped. The path of the balls is impeded by a regular array of pegs, nailed into the backboard. As a ball hits a peg, it bounces to the left or the right with equal probability. Then it hits another peg, and again bounces left or right. The ball continues in this manner until it reaches the bottom.

The Galton board creates a series of perfectly random events, in which the ball bounces left or right with equal probability. When you drop in one ball, you have no idea where it will end up. But when you drop in a large number of balls, you can predict with remarkable accuracy where most of them will land. In fact, you can predict the shape of the curve that the pile of balls will trace out. As Galton discovered, you get a bell curve.

Galton had his answer. Regression to the mean was caused by random effects, which outweighed any tendency for parents to pass on their 'special' (i.e., far from the mean) characteristics to their offspring. Similarly, random variations in performance lead to regression to the mean in the fortunes of football teams and baseball batters, in the performance of symphony orchestras, and in the weather.

The abstract patterns of chance

Today, statisticians have developed the art and science of collecting data to the point at which often highly reliable predictions can be made—predictions of future events based on what has happened in the past, and predictions of the properties of entire populations based on data from a small sample. In so doing, they have extended to the otherwise highly unpredictable living world some of the precision that mathematicians of the sixteenth and seventeenth centuries brought to games of chance.

At the same time, pure mathematicians have been examining the very patterns of probability theory itself, seeking to write down axioms for probability theory much as Euclid wrote down axioms for geometry.

The first axiomatization that was generally accepted was produced by the Russian mathematician A. N. Kolmogorov in the early 1930s. Kolmogorov defined probability as a function from sets to real numbers between 0 and 1. The collection of sets for which this function is defined has to constitute what is called a field of sets. This means that all the sets are subsets of some single set U, itself in the collection; the union and intersection of any two sets in the collection are also in the collection; and the complement (with respect to U) of any set in the collection is also in the collection.

To be a probability function, Kolmogorov required that a function satisfy two conditions: First, the value assigned to the empty set is 0 and the probability assigned to U is 1. Second, if two sets in the field have no elements in common, then the probability of their union is the sum of the probabilities of the two sets.

At the degree of abstraction achieved by Kolmogorov, it became clear that probability theory has much in common with another area of mathematics: *measure theory*. Measure theory had been introduced by Émile Borel, Henri-Léon Lebesgue, and others as a highly abstract study of area and volume. Their motivation was to understand and generalize the integral calculus that Newton and Leibniz had developed in the seventeenth century. Thus, a mathematical field of study that began as an attempt to help gamblers win money in the casino turned out to be fundamentally the same as a field that studies the motion and change we see in the world and the universe—a striking illustration of the incredible power of mathematical abstraction.

Let's take a look at just one way in which calculus and probability theory can help us to make decisions about future events.

Using mathematics to choose your options

In 1997, the Nobel Prize in economics was shared by Stanford University professor of finance (emeritus) Myron Scholes and economist Robert C. Merton of Harvard University. The prize would undoubtedly have been shared with a third person, Fischer Black, but for the latter's untimely death in 1995. The prize was awarded for the discovery, in 1970, of a single mathematical formula: the Black–Scholes formula.

Discovered by Scholes and Black, and developed by Merton, the Black–Scholes formula tells investors what value to put on a financial derivative. A derivative is a financial instrument that has no value of its own; rather, it *derives* its value from the value of some other asset. A common example is the stock option, which provides the purchaser with the option—but not the obligation—to purchase stock at an agreed-upon price before a specified date. Derivatives amount to 'side bets' used to offset the risks associated with doing business under the ever-changing circumstances of the everyday world.

The Black–Scholes formula provides a means for determining the price to be put on a particular derivative. By turning what would otherwise be a guessing game into a mathematical science, the Black–Scholes formula made the derivatives market into the hugely lucrative industry it is today.

So revolutionary was the very idea that you could use mathematics to price derivatives that initially Black and Scholes had difficulty publishing their work. When they first tried in 1970, the University of Chicago's *Journal of Political Economy* and Harvard's *Review of Economics and Statistics* both rejected the paper without even bothering to have it reviewed. It was only in 1973, after some influential members of the

Chicago faculty put pressure on the editors, that the *Journal of Political Economy* published the paper.

Industry was far less shortsighted than the denizens of the ivory tower. Within six months of the publication of the Black–Scholes article, Texas Instruments had incorporated the new formula into its latest calculator, announcing the new feature with a half-page ad in the *Wall Street Journal*.

Modern risk management, including insurance, stock trading, and investment, rests upon the fact that you can use mathematics to predict the future. Not with 100 percent accuracy, of course, but well enough so that you can make a wise decision as to where to put your money. In essence, when you take out insurance or purchase stock, the real commodity you are dealing in is risk. The underlying ethos in the financial markets is that the more risk you are prepared to take, the greater the potential reward. Using mathematics can never remove the risk. But it can tell you just how much of a risk you are taking, and help you decide on a fair price.

What Black and Scholes did was to find a way of determining the fair price to charge for a derivative such as a stock option. Here's how stock options work. You purchase an option to buy stock at an agreed-upon price before some fixed later date. If the value of the stock rises above the agreed-upon price before the option runs out, you buy the stock at the agreed-upon price and thereby make a profit. If you want, you can simply sell the stock immediately and realize your profit. If the stock does not rise above the agreed-upon price, then you don't have to buy it, but you lose the money you paid to purchase the option in the first place.

What makes stock options attractive is that the purchaser knows in advance what the maximum loss is: the cost of the option. The potential profit is theoretically limitless: if the stock value rises dramatically before the option runs out, you stand to make a killing. Stock options are particularly attractive when they are for stock in a market that experiences large, rapid fluctuations, such as the computer and software industries.

The question is, how do you decide on a fair price to charge for an option on a particular stock? This is precisely the question that Scholes, Black, and Merton investigated back in the late 1960s. Black was a mathematical physicist with a recent doctorate from Harvard, who had left physics and was working for Arthur D. Little, the Boston-based management consulting firm. Scholes had just obtained a Ph.D. in finance from the University of Chicago. Merton had obtained a bachelor of science degree in mathematical engineering at New York's Columbia

University, and had found a job as a teaching assistant in economics at M.I.T.

The three young researchers—all were still in their twenties—set about trying to find an answer using mathematics, exactly the way a physicist or an engineer approaches a problem. After all, Pascal and Fermat had shown that you can use mathematics to determine the fair price for a bet on some future event, and gamblers ever since had used mathematics to figure their best odds in games of chance. Similarly, actuaries used mathematics to determine the right premium to charge on an insurance policy, which is also a bet on what will or will not happen in the future.

But would a mathematical approach work in the new, highly volatile world of options trading, which was just being developed at the time? (The Chicago Options Exchange opened in April 1973, just one month before the Black–Scholes paper appeared in print.) Many senior marketeers thought that such an approach couldn't possibly work, and that options trading was beyond mathematics. If that were the case, then options trading was an entirely wild gamble, strictly for the foolhardy.

The old guard were wrong. Mathematics could be applied to the pricing of options—in this case, a mixture of probability theory and calculus known as *stochastic differential equations.* The Black–Scholes formula takes four input variables—duration of the option, prices, interest rates, and market volatility—and produces a price that should be charged for the option.

Not only did the new formula work, it transformed the market. When the Chicago Options Exchange first opened in 1973, fewer than 1,000 options were traded on the first day. By 1995, over a million options were changing hands each day.

So great was the role played by the Black–Scholes formula (and extensions due to Merton) in the growth of the new options market that, when the American stock market crashed in 1978, the influential business magazine *Forbes* put the blame squarely on that one formula. Scholes himself has said that it was not so much the formula that was to blame, but rather that marketeers had not grown sufficiently sophisticated in how to use it.

The awarding of a Nobel Prize to Scholes and Merton showed that the entire world now recognized the significant effect on our lives that had been wrought by the discovery of that one mathematical formula, and emphasized once again the way in which mathematics can transform the way we live.

CHAPTER 8

Uncovering the Hidden Patterns of the Universe

The wanderers

For all our late-twentieth-century sophistication, there are few of us who can gaze up at the sky on a particularly clear night and not be filled with a sense of awe. Knowledge that the twinkling bright lights we see are nature's own nuclear furnaces, each one a sun much like our own, does not diminish the sheer spectacle of it all. Knowing, as we now do, that the light from many of the stars we see has taken millions of years to reach us only enhances our overwhelming sense of the sheer magnitude of the universe in which we find ourselves.

When we consider our own reaction, it comes as no surprise to discover that our ancestors felt a similar sense of wonderment, and to learn that some of the first attempts to understand patterns of nature were directed at the stars.

The ancient Egyptians and Babylonians made observations of the sun and the moon, and used their knowledge of the regular movements of both to construct a calendar and to monitor the seasons in order to guide their agriculture. But both civilizations lacked the mathematical skill to develop a coherent theory of the heavenly bodies they observed. That key step was taken by the Greeks around 600 B.C. Both Thales (whom we met earlier as the person generally credited with the introduction of the idea of proof in mathematics) and Pythagoras appear to have made serious attempts to understand—using mathematics—the complicated motions of some of the stars. In fact, we now know that the 'stars' whose complicated motions perplexed them the most were not stars at all, but rather the planets of our own solar system. The very word 'planet' comes from the complex motions the Greeks observed: the Greek word *planet* means 'wanderer'.

The Pythagoreans declared that the earth must be a sphere, a view that, for all the lack of supporting evidence, gradually came to be accepted by the other Greek thinkers. Mathematical confirmation of the earth's spherical nature was finally provided by Eratosthenes around 250 B.C. By taking measurements of the sun's elevation at two different locations, Eratosthenes not only provided strong confirming evidence of the earth's spherical shape, but also calculated its diameter with what we now know to be 99 percent accuracy.

Prior to Eratosthenes, Plato's student Eudoxus (ca. 408–355 B.C.) proposed a model of the universe as a series of concentric spheres with Earth at its immovable center, with each star moving on one of the spheres. It is not clear how Eudoxus explained the complex motions of the planets (the wandering stars), and his own writings on the subject have not survived. It is also unclear how Eudoxus addressed the fact that the brightness of the planets varies over time. If each planet moved on a fixed sphere around an immovable Earth at the center, the planet's brightness should surely remain constant. But for all its technical shortcomings, Eudoxus' theory is notable for its attempt to provide a mathematical framework to describe the heavens.

In fact, in the middle of the fifth century B.C., Heraclitus had provided a possible explanation for both the varying brightness of the planets and their complex motions by putting forward two revolutionary suggestions: first, that Earth rotated on its own axis, and second, that the peculiar motions of Venus and Mercury were caused by their moving in circles whose center was the Sun. Some time after Eudoxus, around 300 B.C., Aristarchus of Samos went one step further than Heraclitus when he conjectured that Earth too moved around the Sun.

Neither suggestion gained much acceptance, and when Hipparchus took up Eudoxus' circle-based approach in 150 B.C. or thereabouts, he assumed once again that Earth must be the immovable center. Where Hipparchus went beyond Eudoxus was in assuming that the complex-looking motion of the planets were caused by their having circular orbits whose centers themselves moved in circles—much as we know today that the Moon orbits around Earth, which in turn orbits around the Sun.

The immovable-Earth-centered view was also adopted by Ptolemy, arguably the greatest of the ancient Greek astronomers, whose thirteen-volume work *Almagest*, written in the second century A.D., was to dominate European astronomy for 1,400 years. Ptolemy's theory took Eudoxus' circle-based approach and turned it into a precise mathematical model that agreed well with the increasingly accurate astronomical measurements that the later Greeks were able to make.

Reducing the number of circles we move in

With intellectual life in the post-Greek period from around A.D. 500 to A.D. 1500 dominated by the Catholic Church, there was little incentive—and considerable disincentive—for anyone to try to provide a scientific explanation of the workings of the universe. According to the prevailing church doctrine, one simply did not question God's actual designs, which were His and His alone. On the other hand, it was part of the church's teaching that man should strive to understand God's will. Seeing a possible loophole in this mandate, a few bold thinkers of the sixteenth century put forward the new doctrine that God had structured the universe according to mathematical laws, in which case it was not only possible, but was God's will, that efforts be made to understand those laws.

With the doorway to the pursuit of mathematical astronomy once again opened, these early Renaissance thinkers were able to resurrect the ancient Greek ideas and buttress them with the much more accurate measurements that were being made. Gradually, mystical speculation based on dogma gave way to reasoned mathematical analysis. The idea that, in carrying out their investigations, the mathematicians and astronomers were simply trying to understand God's will protected them until those investigations led to a direct confrontation with one of the most fundamental of the church's tenets: the central position of Earth in the universe. The luckless messenger who found himself at the center of the storm was Nicolas Copernicus, who lived from 1473 to 1543.

By the time Copernicus arrived on the scene, attempts to modify the Ptolemaic model to fit the growing masses of data had led to a complicated system using some seventy-seven circles to describe the motions of the Sun, the Moon, and the five planets then known. Knowing that Aristarchus and other Greeks had proposed a heliocentric (or Sun-centered) model of the universe, Copernicus asked himself whether a simpler explanation could be obtained if the Sun were taken to be at the center instead of Earth. Though it still required some ingenuity to produce a model that accounted for the known data, he was able to reduce the seventy-seven circles needed for an Earth-centered model to thirty-four circles if the Sun was at the center.

On mathematical grounds, the heliocentric Copernican model was far superior to its predecessor, but it drew strong opposition from the church, which called the new theory "more odious and more harmful to Christendom than anything whatsoever contained in Calvin's and Luther's writings or in all other heresies." The Danish astronomer Tycho Brahe took enormous pains to obtain accurate astronomical measurements that he thought would help him overturn Copernicus's theory, but in the end the additional data he obtained simply confirmed the superiority of the heliocentric model.

In fact, it was largely based on Brahe's detailed and extensive measurements that his collaborator Johannes Kepler (1571–1630) was able to conclude that the planets moved around the Sun in elliptical orbits, not circular ones. The three laws of planetary motion that Kepler formulated (which we have met already in Chapter 4) provide an excellent example of the scientific precision that was to become typical of the period following the scientific revolution of which Kepler was a part:

1. The planets move around the sun in elliptical orbits with the Sun as one of the two foci.
2. As a planet moves around its orbit, the areas swept out by the radius vector in equal time intervals are equal.
3. The squares of the planets' periods of revolution around the Sun vary as the cubes of their mean distances from the Sun.

By capturing the pattern of planetary motion in such a simple fashion, Kepler's model was far superior to Copernicus's circle-based theory. Moreover, with Kepler's laws, it became possible to predict, with relative ease, the exact position of a planet in the sky at any time in the future.

Final confirmation of Kepler's model—and the final nail in the coffin of the Earth-centered model mandated by the church—was obtained when Galileo Galilei used the newly invented telescope to make extremely accurate measurements of the planets.

Though Galileo's works were banned by the church from 1633 to 1822, during which time the 'official' position of the church was still that Earth was the center of the universe, practically all post-Renaissance scientists had no doubt that Copernicus's heliocentric model was the correct one. What made them so certain? After all, the evidence provided by our everyday experience tells us that the earth beneath our feet is stationary and the Sun, the Moon, and the stars move across the sky above us.

The answer is mathematics: the only reason for adopting the heliocentric model was that the mathematics is much simpler with the Sun at the center of the solar system than with Earth at the center. The Copernican revolution marked the first time in history when mathematics—or more precisely, the desire for a simpler mathematical explanation—forced people to reject the evidence of their own eyes.

The man who made numbers count

The confirmation of Kepler's laws was just one of the major scientific achievements of Galileo Galilei. Born in Florence, Italy, in 1564, Galileo entered the University of Pisa at age seventeen to study medicine, but reading the works of Euclid and Aristotle persuaded him to turn his attention to science and mathematics. It was to be a highly significant change of direction, not just for Galileo himself, but for all humankind. For, together with his contemporary René Descartes, Galileo began the scientific revolution that led directly to present-day science and technology.

While Descartes emphasized the importance of logical reasoning based on empirical evidence, Galileo placed his emphasis on measurement. In so doing, Galileo changed the nature of science. Instead of seeking the underlying *causes* of various natural phenomena, which had been the purpose of science (insofar as there was such a thing) since the time of the ancient Greeks, Galileo sought numerical relationships between different measured quantities. For example, instead of trying to find an explanation of what causes an object to fall to the ground when released from the top of a tower, Galileo tried to find out how the position of the object varies with the time since it was dropped. To do this, he dropped small, heavy balls from a high position (often said to be the top of the Leaning Tower of Pisa, though there is no evidence for this) and measured the time it took them to reach various points on the way down. He discovered that the distance traveled by a ball at any instant varies with the square of the time of fall. Using modern algebraic terminology, he discovered the relationship $d = kt^2$ that connects the distance of fall d with the time of fall t, where k is a constant.

Today, we are so used to mathematical laws of this kind that it is easy to forget that this way of looking at natural phenomena is only four hundred years old, and is highly artificial. In order to obtain a mathematical formula like Galileo's, you have to identify certain features of the world that can be measured, and look for meaningful relationships between those features. The kinds of features for which this approach works are time, length, area, volume, weight, speed, acceleration, inertia, force, moment, and temperature. Features that have to be ignored are color, texture, smell, and taste. If you stop and think about it for a moment, you will realize that most of the features in the first list, the ones to which Galileo's approach can be applied, are purely *mathematical inventions*—they make sense only in terms of assigning numbers to different phenomena. (Even the ones that do have a nonnumerical sense have to be made numerical in order to apply Galileo's approach.)

A mathematical formula relates two or more of these numerical features, and in so doing provides some kind of *description* of the phenomenon of interest, but it does not give an *explanation* of that phenomenon—it does not tell you the *cause*. This was a truly revolutionary way to 'do science', and initially it met with considerable opposition. Even Descartes was skeptical, saying at one point, "Everything Galileo says about bodies falling in empty space is built without foundation; he ought first to have determined the nature of weight." But the genius of the new approach was apparent in its enormous success. Most of the 'patterns of nature' that are now studied using mathematics are patterns that arise in Galileo's invisible, quantitative universe.

How apples fall

Galileo died in 1642. That very same year, Isaac Newton was born in the English village of Woolstorp (see Chapter 3). Newton was one of the first scientists to fully embrace Galileo's new quantitative approach to science, and in Newton's theory of force and gravity we find a dramatic illustration of the artificial, abstract nature of that approach. Consider Newton's well-known law of force:

The total force on a body is the product of its mass and its acceleration.

This law provides an exact relationship between three highly abstract phenomena: force, mass, and acceleration. It is often expressed as a mathematical equation,

$F = m \times a$.

The use of boldface in this equation is the standard mathematician's way of indicating a *vector quantity*, one that has direction as well as magnitude. In Newton's equation, both force F and acceleration a are vectors. As expressed in this equation, force and acceleration are purely mathematical concepts, and on close examination, mass too turns out to have far less 'physical' reality than we might have expected.

For another example, take Newton's famous inverse square law of gravity:

The gravitational attraction between two bodies is proportional to the product of their masses divided by the square of the distance between them.

Written in the form of an equation, the law looks like this:

$$F = k \times \frac{M \times m}{r^2}$$

where F denotes the magnitude of the gravitational force, M and m are the two masses, and r is the distance between them.

Newton's law of gravity has proved to be extremely useful. For instance, in 1820, astronomers observed some inexplicable variations in the orbit of the planet Uranus. The most likely explanation was that the orbit was being affected by the gravitational pull of some hitherto unknown planet. The unknown planet was given the name Neptune. But did it really exist?

In 1841, the English astronomer John Couch Adams carried out a detailed mathematical analysis of Uranus's orbit, using Newton's law, and was able to calculate both the mass and the precise orbit of the unknown planet. Though Adams's results were initially ignored, in 1846 the German astronomer Galle obtained them, and within a matter of hours at the telescope, he had found Neptune. Given the relatively primitive nature of telescopes in those days, it is highly unlikely that Neptune would have been discovered without the exact calculations that Newton's law had made possible.

For a more contemporary example, Newton's law forms the basis of the control systems we use to put communications satellites into stationary orbit around the earth and to send both manned and unmanned spacecraft on long interplanetary voyages.

And yet, for all that Newton's law of gravity is highly precise, it tells us nothing about the *nature* of gravity—about what it *is*. The law provides a mathematical description of gravity. To this day, we do not have a physical description. (Thanks to Einstein, we do have a *partly* physical explanation of gravity as a manifestation of the curvature of space–time. But the mathematics required to handle Einstein's explanation is considerably more complex than the simple algebra of Newton's law.)

The invisible threads that bring us together

These days, we think nothing of having a telephone conversation with someone on the other side of the world, of seeing live television pictures of an event taking place many miles, perhaps many thousands of miles, away, or of using a radio to listen to music being played by a person in a studio on the other side of town. As inhabitants of Planet Earth, we have been drawn closer together by our modern communications technologies, which allow us to transmit information over long distances by means of invisible 'waves', called radio waves or, more technically, *electromagnetic* waves. (The reason for the quotation marks around the word 'waves' will become clear in due course.)

In fact, the rapid growth of electromagnetically mediated communications technologies over the past thirty years almost certainly marks a fourth major change in the very nature of human life. Our acquisition of language some hundred thousand years or so ago occasioned the first such major shift, allowing our early ancestors to pass on large amounts of information from person to person and from generation to generation.

With the invention of written language about seven thousand years ago, humankind was able to create relatively permanent records of information, enabling people to use written materials as a means of communication over sometimes great distances and long time periods.

The invention of the printing press in the sixteenth century provided the means for one person to *broadcast* information simultaneously to a large number of other people. Each of these developments served to bring people socially closer together and enabled them to collaborate on greater and greater projects.

Finally, with the introduction of telephone, radio, television, and the more recent forms of electronic communication, our human society seems to be developing rapidly into one in which so many people can coordinate their activities to such an extent that a genuine collective intelligence is achieved. To take just one example, without the ability to bring many thousands of people together into a single team, working largely in unison, there is no way we could have sent a man to the moon and brought him back alive.

Modern life is quite literally held together by the invisible electromagnetic waves that speed past us—and through us—all the time. We know those waves are there (or at least that something is there) because we see and make use of their effects. But what are they exactly? And how are we able to use them with such precision, and to such great effect? To the first question, we do not yet have an answer. As with gravity, we do not really know what electromagnetic radiation *is*. But there is no doubt what it is that enables us to make such effective use of it: mathematics.

It is mathematics that allows us to 'see' electromagnetic waves, to create them, to control them, and to make use of them. Mathematics provides the only description we have of electromagnetic waves. Indeed, as far as we know, they are *waves* only in the sense that the mathematics treats them as waves. In other words, the mathematics we use to handle the phenomenon we call electromagnetic radiation is a theory of wave motion. Our only justification for using that mathematics is that it works. Whether the actual phenomenon really does consist of waves in some medium is not known. What evidence there is suggests that the wave picture is at best an approximate view of a phenomenon that we may never be able to fully comprehend.

Given the ubiquity of present-day communications technologies, it is hard to appreciate that the scientific understanding upon which they rest is less than a hundred and fifty years old. The first radio signal was sent and received in 1887. The science on which it depended had been discovered just twenty-five years earlier.

Building the Maxwell house

It is often said that today's communications technologies have already turned the world into a single village—a global village—and that we will soon be living as a single family in a global house. Given the social problems and the never-ending conflicts of contemporary society, that metaphor might apply only if the family is highly dysfunctional! But with the availability of instant, real-time communication—both sound and pictures—regardless of where we are in the world, from a social and communications viewpoint, our world really is starting to resemble a house. The science needed to build that house was developed by the English mathematician James Clerk Maxwell, who lived from 1831 to 1879.

The key observation that led Maxwell to his electromagnetic theory was made by accident in 1820. Working in his laboratory one day, the Danish physicist Hans Christian Oersted noticed that a magnetic needle moved when an electric current was passed through a wire in its vicinity. It is often reported that when Oersted told his assistant of his observation, the latter simply shrugged and commented that it happened all the time. Whether or not such an exchange actually took place, Oersted himself felt the phenomenon to be of sufficient interest to report it to the Royal Danish Academy of Sciences. It was the first demonstration that there is a connection between magnetism and electricity.

A further connection between magnetism and electricity was established the following year, when the Frenchman André-Marie Ampère observed that when electric current passes through two wires strung loosely in parallel, close together, the wires behave like magnets. If the current passes through the two wires in the same direction, the wires attract each other; if the current flows in opposite directions, the wires repel each other.

Ten years later, in 1831, the English bookbinder Michael Faraday and the American schoolmaster Joseph Henry independently discovered what is essentially the opposite effect: if a wire coil is subjected to an alternating magnetic field, an electric current is induced in the coil.

It was at this point that Maxwell entered the picture. Starting around 1850, Maxwell sought a scientific theory that would explain the strange connection between the invisible phenomena of magnetism and electricity. He was strongly influenced by the great English physicist William Thomson (Lord Kelvin), who advocated finding a mechanistic explanation of these phenomena. Specifically, having developed a mathematical theory of wave motion in fluids, Thomson suggested that it might be possible to explain magnetism and electricity as some kind of *force fields* in the ether. The *ether* was a postulated—but hitherto undetected—substrate through which heat and light traveled.

The concept of a force field, or more simply, a field, is highly abstract and can be described only in a purely mathematical way (as a mathematical object called a *vector field*). You can obtain a view of the 'lines of force' in a magnetic force field by placing a card on top of a magnet and sprinkling iron filings onto the card. When the card is tapped lightly, the filings align themselves into an elegant pattern of curved lines that follow the invisible lines of magnetic force (see Figure 8.1).

A similar 'view' of an electric force field can be obtained by passing electric current through a wire that passes through a hole in a card, sprinkling iron filings on the card, and again tapping the card lightly. The filings will arrange themselves into a series of concentric circles centered at the wire (see Figure 8.2).

To the mathematician, a force field is a region in which there is a force acting at each point. Both the magnitude and the direction of the force vary with your position as you move around the region, generally in a continuous, unbroken fashion. If you were to move about in such a field, the force you would be subjected to would vary in both magnitude and direction. In many force fields, the force at each point also varies with time.

Recognizing that the only way to explain something as abstract as a force field was in a Galilean fashion, using mathematics, Maxwell sought to formulate a collection of mathematical equations that would accurately describe the behavior of magnetism and electricity. He was spectacularly successful, publishing his results in 1865 under the title *A Dynamical Theory of the Electromagnetic Field.*

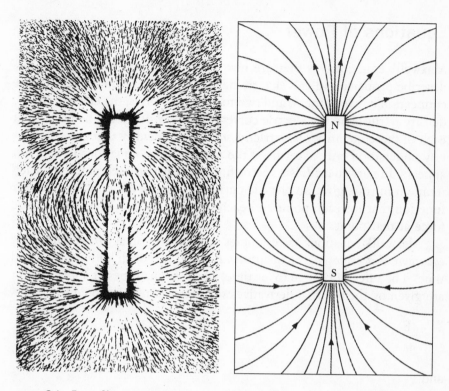

FIGURE **8.1** Iron filings tossed onto a card on top of a magnet align themselves with the lines of force of the magnetic field.

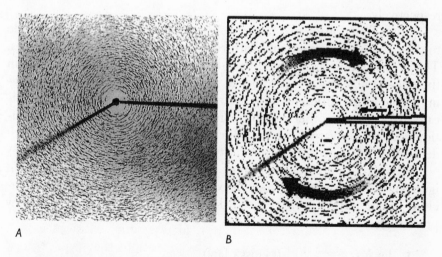

A

B

FIGURE **8.2** Iron filings tossed onto a card surrounding a wire carrying an electric current align themselves with the lines of force of the magnetic field generated by the current.

Equations of Power

Maxwell's equations, as they are now known, are four in number. They describe properties of the electric field E and the magnetic field B, the connection between them, and two further quantities, the electric charge density ρ (rho) and the electric current density j. It should be noted that each of these quantities is a strictly *mathematical* entity that can be properly understood only in terms of the mathematical equations in which it appears. In particular, E is a vector function that for each point and each moment in time gives an electric current at that point, and B is a vector function that for each point and each moment in time gives a magnetic force at that point.

Maxwell's equations are examples of what are called *partial differential equations*. For example, the quantities E and B vary with both position and time. The rates of change (over time) of these quantities at a given point are given by their partial derivatives with respect to t, denoted by

$$\frac{\partial E}{\partial t}$$

and

$$\frac{\partial B}{\partial t},$$

respectively.

[For a function of two (or more) variables, such as E, it is possible to define two derivatives, one with respect to each of the two variables. The actual definition of these two derivatives is exactly the same as in the one-variable case we considered in Chapter 3; the variable not playing a role in the differentiation is simply treated as a constant. To indicate that there are other variables, mathematicians use a different notation for the derivative, writing, for example, $\frac{\partial E}{\partial t}$ instead of $\frac{\partial B}{\partial t}$. Such derivatives are called *partial derivatives*.]

Taking the simplest case, that of electromagnetism in a vacuum, and ignoring some constant factors, Maxwell's four equations are (and I include them purely for the record—explanations in English follow presently):

1. div $E = \rho$ (Gauss's law)

2. **curl** $B = j + \dfrac{\partial E}{\partial t}$ (Ampère–Maxwell law)

3. curl $E = -\dfrac{\partial B}{\partial t}$ (Faraday's law)

4. div $B = 0$ (No magnetic monopoles)

The operations div (divergence) and **curl** are operations from vector calculus—the calculus you get when you try to compute rates of change in a field. For any field F, the flux of F from a given volume is a measure of the number of field lines that emerge from that volume—roughly, how intense is the force F around that volume, or how much of F is flowing out of that volume. The quantity div F gives, at each point in the field, the flux of the field per unit of volume that flows out through a small sphere around the point. The curl of F at a point measures the rotation vector for a spatial vortex in a small neighborhood of that point—roughly, how much the direction of the field is turning around in the region of that point.

In descriptive (i.e., nonquantitative), verbal terms, Maxwell's equations can be expressed, somewhat loosely, as follows:

1. The flux of the electric field out of a finite volume is proportional to the electric charge within that volume.
2. An electric current or an alternating electric flux gives rise to a magnetic eddy.
3. An alternating magnetic flux gives rise to an electric eddy.
4. The total magnetic flux from a finite volume is always zero.

Maxwell's equations imply that if an electric current is allowed to fluctuate back and forth in a conductor (such as a wire), the resulting electromagnetic field, which alternates in time with the current, will tear itself free from the conductor and flow into space in the form of an *electromagnetic wave*. The frequency of the wave will be the same as that of the current that causes it. (This is the basis for radio and television transmission.) Using his equations, Maxwell was able to calculate the speed of the electromagnetic wave that breaks free: about 300,000 kilometers per second (about 186,000 miles per second).

In what sense is this rapidly moving entity a wave? Strictly speaking, what the mathematics gives you is simply a mathematical function—a solution to Maxwell's equations. It is, however, the same kind of function that arises when you study, say, wave motion in a gas or a liquid. Thus, it makes perfect mathematical sense to refer to it as a wave. But remember, when we are working with Maxwell's equations, we are working in a Galilean mathematical world of our own creation. The relationships between the different mathematical entities in our equations will (if we have set things up correctly) correspond extremely well to the corresponding features of the real-world phenomenon we are trying to study. Thus, our mathematics will give us what might turn out to be an extremely useful *description*—but it will not provide us with a true *explanation*.

Math light

Wave or not, as Maxwell noticed with excitement, there was something very familiar about that speed of propagation he had computed. It was very close to the speed of light, which was known with reasonable accuracy at the time—close enough that the two speeds might actually be the same. (One early calculation of the speed of light, carried out in 1673, depended on the observation that eclipses of Jupiter's moon Io were about 16 minutes later when observed while Earth was farthest away from Jupiter than while Earth was closest to Jupiter. Assuming that the delay is caused by the extra distance the light has to travel—roughly the diameter of Earth's orbit, approximately 300 million kilometers— that gives a figure of roughly 312,000 kilometers per second for the speed of light. More precise measurements made in the late nineteenth century gave a figure of just under 300,000 kilometers per second.)

In fact, Maxwell thought, maybe light and electromagnetic radiation don't just travel at the same speed. Perhaps they are one and the same, he suggested in 1862. Maybe light is just a particular form of electromagnetic radiation—say, electromagnetic radiation of a particular frequency?

At the time, there were two conflicting theories as to the nature of light. One theory, the corpuscle theory, proposed by Newton around 1650, held that light consists of small invisible particles, called corpuscles, that are radiated by all luminous bodies and which travel in straight lines. The other theory, that light consists of a wave, was put forward by Christiaan Huygens at about the same time. (We now know that neither the corpuscle theory nor the wave theory is completely 'correct', in that each theory accords with some phenomena far better than does the other.)

By Maxwell's time, the wave theory was favored by most physicists, so Maxwell's suggestion that light was simply a form of electromagnetic radiation should have been well received. But it wasn't. The problem was the highly mathematical nature of Maxwell's theory. For instance, in a speech given in 1884, Lord Kelvin declared that Maxwell's work did not provide a satisfactory account of light, and that physicists should continue the search for an explanatory, mechanistic model. Maxwell himself was certainly not immune to these criticisms, and made a number of attempts to provide a mechanistic explanation, but none were successful—and indeed, to this day, we have no such explanation.

But for all its lack of an intuitive explanation, Maxwell's theory was both scientifically powerful and extremely useful, and today it is accepted as the standard mathematical description of electromagnetic radiation. As early as 1887, the German physicist Heinrich Hertz successfully generated an electromagnetic wave in one circuit and received it in a second circuit some distance away. In simple terms, he made the world's first radio transmission. A few years later, radio waves were being used to transmit human voices over increasingly long distances, and a mere eighty-two years after Hertz's experiment, in 1969, a human being standing on the surface of the moon used radio waves to communicate with his colleagues back on Earth.

Today, we know that light is indeed a form of electromagnetic radiation, and that there is a whole range of electromagnetic radiation, which spans a spectrum according to wavelength, ranging from 10^{-14} meters at the high-frequency end to 10^8 meters at the low end. The radio waves used to transmit radio and television signals have much lower frequencies than light waves, and lie at the lower end of the electromagnetic spectrum.

At higher frequencies than radio waves but lower frequencies than light are the infrared waves, which are invisible, but transmit heat. Light itself constitutes the visible portion of the electromagnetic spectrum, with red at the lower-frequency end and violet at the higher-frequency end, and the familiar colors of the rainbow—orange, yellow, green, blue, and indigo—arranged in order in between. Radiation with a frequency slightly higher than violet is known as ultraviolet radiation. Though not visible to the human eye, it will blacken a photographic film and can be seen using special equipment.

Beyond the ultraviolet region we find the invisible X-rays, capable not only of blackening a photographic film but also of penetrating human flesh, a combination of features that has led to their widespread use in medicine.

Finally, at the uppermost end of the spectrum are the gamma rays emitted by radioactive substances as they decay. In recent years, gamma rays, too, have found uses in medicine.

Human vision, communication, medicine—even microwave cooking—all make extensive use of electromagnetic radiation. They all depend on Maxwell's theory, and serve as a testament to the accuracy and the precision offered by his four equations. And yet, as we have already observed, they don't provide us with an *explanation* of electromagnetic radiation. The theory is purely mathematical. It provides yet another dramatic illustration of the way in which mathematics enables us to 'see the invisible'.

Gone with the wind

One obvious question left unanswered by Maxwell's electromagnetic theory was: what is the nature of the medium through which electromagnetic waves travel? Physicists called this unknown medium the ether, but had no idea of its nature. Spurred on by the scientist's usual desire for as simple a theory as possible, they assumed that the mysterious ether was at rest everywhere in the universe, a constant backdrop against which the stars and the planets moved and through which light and other electromagnetic waves rippled.

To test this assumption of a fixed ether, in 1881, the American physicist Albert Michelson devised an ingenious experiment to try to detect it. If the ether is at rest and the earth moves through it, he proposed, then from the standpoint of an observer on the earth, there would be an 'ether wind' blowing steadily in the direction opposite to the earth's motion. Michelson set out to detect this wind, or more precisely, its effect on the velocity of light. His idea was to generate two light signals at the same instant, one in the direction of the earth's supposed motion through the ether, the other perpendicular to it. To do this, he generated a single light signal and used a half-silvered mirror at 45° to the light beam to split the signal into two perpendicular beams. The two beams were directed toward two other mirrors placed exactly the same distance from the splitting mirror, which reflected the two beams back to a detector. The entire apparatus is illustrated in Figure 8.3.

Because one of the two beams travels first into the face of the supposed ether wind and then, after reflection, travels back with the wind,

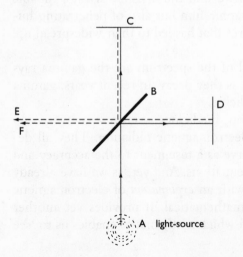

FIGURE **8.3** Michelson's experiment. A light beam from the source (A) is split by a half-silvered glass plate (B). The two resulting beams are reflected back by mirrors C and D, merging to give the beam comprising waves E and F. The distances BC and BD are equal. Michelson observed no interference between the two waves E and F.

whereas the other beam travels perpendicular to the ether wind, there should be a slight discrepancy in the times when the two beams reach the detector: the beam that first travels into the wind should arrive slightly later than the other. Imagine two equally matched swimmers, one who swims into the current and then back to the starting point, the other who swims an equal distance out and back perpendicular to the current. The first swimmer will arrive back after the second. In fact, if she swims perpendicular to the current the whole time, the second swimmer will not arrive back exactly at the starting point, but will be carried along in the direction of the current. The same would be true of the light wave in Michelson's experiment, but because of the high speed of light—about 300,000 kilometers per second—compared with the supposed speed of the earth moving though the ether—the orbital speed of the earth is about 30 kilometers per second—the drift will be negligible, and not detectable by the receiving apparatus.

Michelson's idea was to measure the discrepancy between the arrival times of the two beams at the detection apparatus. But how? Even today, we don't have clocks that could measure the tiny difference in time that would result from the effect of a 30-kilometer-per-second 'ether wind' on a wave traveling at 300,000 kilometers per second. Michelson's brilliant solution to this conundrum was to use the light itself to detect any difference. Because they originated from the same source, when the two beams left the splitting mirror, they would consist of light waves exactly in synchronization, crest for crest. If they traveled at different speeds, then when they returned to the detector, they would be out of synchronization, and the resulting combined wave would be different from the original one sent out—a difference that could, quite literally, be *observed*.

However, contrary to everyone's hopes and expectations, Michelson detected no alteration in the returning, combined wave. The two beams were still in perfect synchronization when they merged after their separate journeys. Not quite believing this result, Michelson repeated the experiment several times over the next few years. One possibility he wanted to eliminate was that the experiment was performed just when Earth's orbit had put it into a position of rest relative to the ether. Another was that the apparatus was angled at exactly 90° relative to the ether wind. However, no matter what time of year he tried, and no matter how he angled the apparatus, he was unable to detect any difference in the time it took the two light beams to reach the detector.

What had gone wrong? It seemed as though there was no ether wind—and hence no ether. But then, what carried electromagnetic waves? To try to escape this dilemma, two physicists, Hendrik Antoon

Lorentz and George FitzGerald, independently came up with a radical suggestion: that when any object traveled through the ether, its length decreased in the direction of ether travel by an amount that, when applied to Michelson's apparatus, would exactly compensate for the difference in the time it would otherwise take for the two electromagnetic beams to reach the detector. This suggestion was not only radical, but seemed highly contrived, and initially it was met with considerable skepticism. But in fact, subsequent developments showed that, however unlikely it seemed, the Lorentz–FitzGerald explanation was not far from the truth.

The resolution of the puzzle of the undetectable ether wind was finally supplied by Albert Einstein with his famous theory of relativity, in 1905.

The most famous scientist of all time

If asked to name a scientist—or even a 'genius'—the average person in the Western world will almost certainly name Albert Einstein. For some reason, the name of this German-born Swiss patent officer turned Princeton academic has become virtually synonymous with scientific genius, at least in common culture.

Born in Ulm in 1879, Einstein spent most of his childhood in Munich, which is where he received his schooling. In 1896, he relinquished his German citizenship because of his dislike of German militarism, and was stateless until obtaining Swiss citizenship in 1901. By then, he had moved to Zurich and had graduated from the Swiss Polytechnic Institute. In January 1902, having failed to gain a teaching position at the Polytechnic, he obtained a position in the Swiss Patent Office in Bern. He was a Technical Expert, Third Class.

Three years later, in 1905, he developed his famous theory of special relativity, a scientific breakthrough that within a few years would make him world famous. In 1909 he resigned his position at the Patent Office to accept an appointment as Extraordinary Professor of Physics at the University of Zurich. Fleeing Europe in 1935 to escape the Nazis, the Jewish Einstein accepted a position at the newly created Institute for Advanced Study in Princeton, New Jersey, where he remained for the rest of his career. He became a U.S. citizen in 1940. He died in Princeton in 1955.

To get an initial understanding of relativity, imagine you are in an airplane, flying at night with the window blinds down, so that you cannot see outside. Assuming there is no turbulence, you are not aware of the fact that the airplane is in motion. You get up from your seat and walk around. The flight attendant pours you a cup of coffee. You idly toss a small bag of peanuts from one hand to the other. Everything seems normal, just as if you were at rest on the ground. And yet you are hurtling through the air at 500 miles an hour. How is it that, when you get up from your seat, you are not thrown to the back of the plane? How is it that the poured coffee and the tossed bag of peanuts do not fly back into your chest?

The answer is that all of these motions—the motion of your body, the coffee, and the peanuts—are relative to the motion of the airplane. The interior of the airplane provides what is, in effect, a stationary background—what physicists call a *frame of reference*—relative to which you, the coffee, and the peanuts move. From the point of view of you or anyone else inside the airplane, everything behaves exactly as it would if the airplane were stationary on the ground. It is only when you open the window blinds and look outside, and see the lights on the ground beneath you flash backward, that you can detect the airplane's motion. You can do so because you are able to compare two frames of reference: the airplane and the ground.

The example of the airplane shows that motion is relative: one thing moves relative to another. What we see as—and think of as—'absolute' motion is motion relative to the frame of reference we are in—and are aware of—at the time. But is there a 'preferred' frame of reference—nature's own frame of reference, if you like? Aristotle thought so: in his view, the earth is at rest, and hence all motion relative to the earth is 'absolute motion'. Copernicus believed that all motion is relative. Newton believed in a fixed 'space' against which everything is either absolutely stationary or else in absolute motion. Lorentz likewise assumed that nature has a preferred, stationary frame of reference against which everything else is either at rest or in motion: namely, the ether. Thus, for Lorentz, motion relative to the ether was absolute motion.

Going one step further, Lorentz suggested that if the mass of a body and the forces between various bodies change with the velocity of those bodies through the ether, as do (he supposed) the dimensions of the bodies (his radical idea mentioned a moment ago, known as the Lorentz contraction), then natural phenomena will appear to occur according to exactly the same laws in a moving framework as in one at rest, provided that the various measurements are made with the framework's own yardsticks. Thus, given two observers, one in motion (relative to the ether) and the other at rest, it will be impossible for either to determine who is in motion and who is at rest.

A particular consequence of Lorentz's proposal was that the mass of a body increases with the velocity of the body. The magnitude of the increase predicted by Lorentz's mathematical analysis is too tiny to be detected for relatively low speeds, but when the velocity approaches the speed of light, the change becomes more significant. In recent times, measurements have been made of the increase in mass of beta particles emitted at high speeds in the decay of radioactive materials, and the results have been in complete accord with Lorentz's predictions.

Building on Lorentz's theory, Einstein went one significant step further. He abandoned the idea of a stationary ether altogether, and simply declared that all motion is relative. According to Einstein, there is no preferred frame of reference. This is Einstein's *principle of special relativity*.

In order to square the mathematics with the principle of special relativity, Einstein had to assume that electromagnetic radiation had a very special—and decidedly counterintuitive—property. Whatever your frame of reference, Einstein said, when you measure the speed of light or any other form of electromagnetic radiation, you will always get the same answer. Thus, for Einstein, the one absolute was not some material *substance* through which electromagnetic waves traveled, but the *speed* of electromagnetic waves themselves.

It's about time

By taking the speed of light to be the same in all frames of reference, Einstein was able to resolve another troublesome question: what does it mean to say that two events happen at the same time? This question of *simultaneity* becomes problematic when the events occur a great distance apart.

For Einstein, time was not absolute, but depended on the frame of reference in which it was measured. Light was the key to simultaneity. To illustrate Einstein's ideas, imagine a train traveling at high speed. At each end of the train is a door, which can be opened by sending a light signal from a source in the middle of the train, exactly halfway between the two doors. (This is an example of a hypothetical 'thought experiment', in which you have to assume that all measurements are perfectly accurate, doors open instantaneously, etc.) Suppose you are a passenger on the train, seated in the middle by the light source. (Thus the moving train is your frame of reference.) What do you see when the light signal is generated? You see the two doors open simultaneously. They open at the same moment because it takes the light signal exactly the same time to reach the two doors.

Now suppose you are not on the train, but standing 50 meters from the track, watching the train go by. The light signal is generated. What do you see next? First the rear door opens, then the front door opens. Here's why. Because the train is in motion (relative to your frame of reference), the rear door moves a small distance forward—toward the light—in between the generation of the light and its arrival at the rear door, and the front door moves a small distance away from the light in between the generation of the light and its arrival at the front door. Thus, as seen from your frame of reference outside the train, the light has to travel farther to reach the front door than it does to reach the rear door. But the speed of light is constant in all frames and in all directions, says Einstein. Hence the light beam reaches the rear door first (from your viewpoint), and the rear door opens first (from your viewpoint).

Thus, from the point of view of an observer on the train, the two doors open at the same moment, but from the point of view of an observer on the ground, the rear door opens before the front door. According to Einstein, then, time is not absolute; it is relevant to the standpoint of the observer.

Accounting for the gravity of the situation

For all its power, Einstein's theory of special relativity applied only when two or more frames of reference moved relative to each other at a constant velocity. Moreover, though the theory of special relativity provided information about the nature of space and time, it said nothing about the two other basic constituents of the universe: mass and gravity. In 1915, Einstein found a way to extend his theory of relativity to take account of both. This new theory is called *the theory of general relativity*.

The basis for this new theory was the *principle of general relativity*: that all phenomena occur the same way in all frames of reference, whether in accelerated motion or not. In general relativity, a natural process affected by gravity would occur if there were no gravity and the whole system were in accelerated motion.

For a particular example of the general relativity principle, consider again your nighttime flight on an airplane with all the window shades pulled down. If the aircraft suddenly accelerates, you will feel a force pushing you toward the back of the plane. Indeed, if you happen to be standing in the aisle at the time, you might find yourself flung toward the rear of the plane. Likewise, when the plane decelerates rapidly (as it does on landing), you are subjected to a force pushing you toward the front of the plane. In both cases, an acceleration/deceleration is perceived by you as a force. Since you are unable to see outside, you cannot observe that the airplane is accelerating or decelerating. If you did not know any better, you might be inclined to try to explain your sudden motion toward the back or the front of the plane as being caused by some mysterious force. You might even decide to call that force 'gravity'.

One of the first pieces of evidence that confirmed Einstein's theory of general relativity concerned an old puzzle about the rather peculiar behavior of the planet Mercury. According to Newton's theory of gravitation, the planets move around the sun in fixed elliptical orbits, in accordance with Kepler's observations. However, the increased accuracy of observations made subsequent to Kepler's time showed that in the case of Mercury, the elliptical orbit was not quite stationary; it fluctuated by a small but detectable amount. At roughly 41 seconds of arc each century, the deviation from the Newtonian prediction was small by everyday human standards, but for scientists it presented a major problem: what caused the deviation?

Until Einstein came along with his theory of general relativity, scientists did not have a plausible explanation. But Einstein's theory of general relativity actually predicted that the orbits of *all* the planets, not just Mercury, should vary. Moreover, the theory gave precise numerical values for the amounts by which the orbits of each of the planets should change. For all the planets except Mercury, the predicted value was too small to be detectable. For Mercury, however, the theoretical value agreed exactly with the figure observed by the astronomers.

The explanation of the orbit of Mercury provided considerable support for Einstein's new theory. But the real clincher came with a dramatic astronomical observation carried out in 1919. One, initially surprising, consequence of the theory of general relativity is that light should behave as if it had mass. In particular, light waves should be subject to gravity. If a light wave passes close by a large mass, such as a star, the gravitational field of the mass should deflect the light wave. The solar eclipse of 1919 gave astronomers the opportunity to test this prediction, and what they found agreed exactly with Einstein's theory.

To make their observations, the astronomers took advantage of a lucky break. By a sheer fluke, the relative sizes of the Sun, the Earth, and the Moon are such that when the Moon lies directly between the Earth and the Sun, the Moon appears exactly the same size as the Sun, and thus, to an observer situated at the appropriate location on the Earth's surface, the Moon can exactly obscure the Sun. Two teams of astronomers were sent to appropriate locations to observe the eclipse. On the day of the eclipse, when the Moon exactly blocked all the Sun's light, the astronomers took measurements of the positions of distant stars in the vicinity of the Sun. Those stars appeared to be in positions far removed from their usual positions, as observed when the Sun was in a different location. More precisely, stars that were known to be in positions directly behind the eclipsed Sun (see Figure 8.4) could be clearly observed through the telescope. This is exactly what Einstein had predicted, based on his theory of relativity. According to the theory, when the light rays from the stars passed close by the Sun, they were bent around the Sun by the Sun's gravitational field. Thus, a star that should have been invisible, hidden behind the Sun, could be seen.

Moreover, the astronomers were able to measure the apparent positions of the 'hidden' stars, and their findings agreed exactly with the theoretical figures Einstein had produced. That was the clincher. Though Newton's theory of gravity and planetary motion was sufficiently accurate for most everyday purposes, such as calculating lunar calendars and drawing up tide tables, when the time came to do accurate astronomical science, Newton was out and Einstein was in.

The geometry of space-time

Though it is probably not apparent from the account I have just given, Einstein's theories of special and general relativity are, at heart, geometric theories—they tell us about the geometric structure of the universe. Moreover, the mathematics of relativity theory is essentially geometry. That geometry is, however, not the usual Euclidean geometry. It is more accurate to say that it is yet another non-Euclidean geometry, in the tradition begun with the development of projective geometry and the non-Euclidean geometries of Gauss–Lobachevsky–Bolyai and of Riemann, which we met in Chapter 4. More precisely, there are two new geometries, one for special relativity, the other for general relativity, with the former geometry a special case of the latter. In fact, with hind-

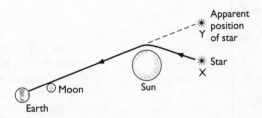

FIGURE **8.4** When light from a distant star passes close to the sun, it is deflected by the sun's gravitational field. This phenomenon was first observed during the solar eclipse in 1919. The star marked X appeared as if it were in the position Y.

sight, it is possible to see how close Riemann came with his non-Euclidean geometry to discovering general relativity before Einstein.

Let's start with special relativity. What does this theory tell us about the geometry of the universe? The first thing to notice is that time and space are closely interconnected. The length of an object changes with its velocity, and simultaneity is closely bound up with the passage of light waves. In the sense of simultaneity, or 'when an event occurs', time travels through space at the speed of light. So, our geometry is a geometry not of the three-dimensional physical universe accessible to our senses, but of the four-dimensional *space–time* universe.

The geometry of special relativity was introduced by Hermann Minkowski, a Russian mathematician who had been one of Einstein's instructors when the latter was a student at the Polytechnic Institute in Zurich. A point in Minkowski space–time has four real-number coordinates, t, x, y, z. The coordinate t is a time coordinate; x, y, and z are space coordinates. When physicists and mathematicians draw space–time diagrams, they usually draw the t-axis vertically upward (so that time flows vertically up the page), and use perspective to draw the x-, y-, and z-axes (or sometimes just two of the three space-axes), as shown in Figure 8.5.

Because of the special—and fundamental—role played by light (or, more generally, by electromagnetic radiation) in relativity, measurement in the t-direction can be thought of either as time or as distance: a time interval of temporal length T can be regarded as a spatial interval of (spatial) length cT, where c is the speed of light. (cT is the distance light trav-

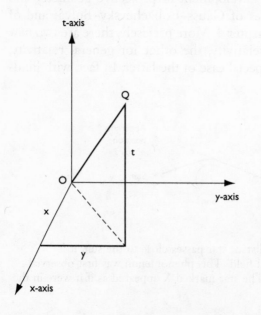

FIGURE **8.5** Minkowski space–time. The time axis is drawn vertically upward. There are three spatial axes, x, y, and z; for clarity, only two of the three are shown here, the x-axis and the y-axis. Any point Q in Minkowski space–time can be described by four coordinates, t, x, y, and z.

els in time T.) In order that measurements along the four axes in Minkowski space–time can be given in the same units, measurement along the t-axis is almost always given in spatial terms, incorporating the multiplier factor c. For instance, we do this in Figure 8.6, which shows a symmetrical four-dimensional double cone having its axis along the t-axis, its center at the origin of coordinates, and its face angled at a slope of 45° to the t-axis. The coordinate equation for the surface of this cone is

$$(ct)^2 = x^2 + y^2 + z^2.$$

This equation is simply a four-dimensional analogue of Pythagoras' theorem. The point to notice is that measurement along the t-axis is in units of c.

The cone shown in Figure 8.6 is particularly important in the geometry of relativity theory. It is called a *light cone*. Imagine a light signal emanating from a point source at O. As time passes, the light wave will radiate outward in an ever-growing sphere. This growing spherical shell in three-dimensional space becomes, in the Minkowski universe, the top half of the light cone (shell). To see this, imagine some moment in time, T, after the light signal has started on its journey. By time T, the light will have traveled a distance cT. The signal will have therefore reached the surface of a sphere in three-dimensional space, centered at O, with radius cT. The equation of such a sphere is

$$(cT)^2 = x^2 + y^2 + z^2,$$

which is a cross-section through the light cone at $t = cT$. Thus, what to an observer living in three-dimensional space appears to be an expand-

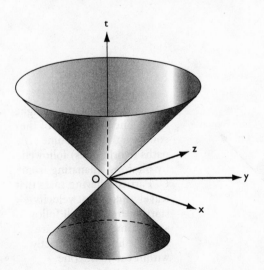

FIGURE **8.6** A light cone in Minkowski space–time.

ing sphere of light would be perceived as the top half of a light cone by an observer outside of time, who sees time as a fourth dimension.

The lower half of the light cone, where $t < 0$, represents the history of a (spherical) light signal that converges on the origin, where it arrives at time $t = 0$. The idea of a shrinking sphere of light is somewhat bizarre—and purely theoretical—and indeed, physicists often ignore the lower, or negative, half of the light cone, which does not correspond to events that we typically encounter. I shall make the same simplification here.

Physicists sometimes think of light as a stream of fundamental particles, called *photons*. With this view of light, the *generators* of the (upper, or positive, half of the) light cone—the straight lines in the surface of the cone through the origin shown in Figure 8.7—represent the paths of individual photons from the light source.

Photons are assumed to have zero mass when at rest. Indeed, according to relativity theory, any particle that can travel at the speed of light has to have zero mass when at rest. (We have to add the caveat 'when at rest' because mass increases with speed.) Particles or objects that have a nonzero mass at rest cannot move at the speed of light; they move more slowly than light. The path through space–time of any particle or object having mass, started in unaccelerated motion at O, will be a straight line through O that lies in the interior of the light cone. Some such paths are also shown in Figure 8.7.

The path of an object through Minkowski space–time is sometimes referred to as the *world line* of the object. Every object has a world line, even an object at rest, for which the world line is the t-axis. (The stationary object endures in time for some period, but its x-, y-, and z-coordinates do not change.)

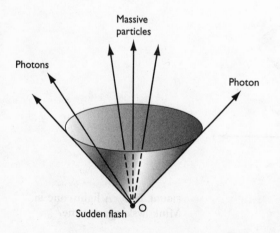

FIGURE **8.7** The upper light cone. The generators of the cone are the space–time paths (world lines) followed by photons emanating from O. Particles having mass that travel at constant velocity starting from O will follow space–time paths that are straight lines lying strictly within the light cone.

An object that starts in motion at O and moves with changing speed will have a world line that is not straight. However, the world line for such an object will lie completely within the light cone, and at all points the angle it makes with the t-axis will be less than 45° (see Figure 8.8).

The interior of the double light cone depicts that part of the universe that is accessible (that *exists*, if you like) to an observer stationed at O. The upper part of the interior of the double light cone, where $t > 0$, represents the future of the universe for an observer stationed at O; the lower part, where $t < 0$, represents the past of the universe for such an observer. To put it a slightly different way, points inside the double cone for which $t < 0$ can be thought of as points in space–time in the observer's (extended) past; points inside the double cone for which $t > 0$ can be thought of as points in space–time in the observer's (extended) future.

The point O itself represents the present moment for the observer. As far as the observer at O is concerned, anything outside the double light cone does not exist—it is outside of the observer's time frame. It is in neither the extended past nor the extended future of the observer.

Going the distance

By representing time as a spatial coordinate, Minkowski's geometry provides us with a visualization of space–time—at least, a partial visualization, limited only by our inability to fully visualize a four-dimensional

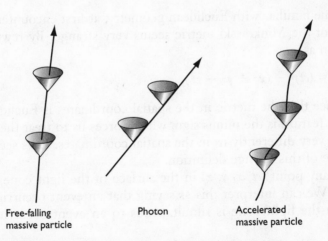

Free-falling massive particle   Photon   Accelerated massive particle

FIGURE **8.8** A massive particle traveling under acceleration will trace out a curved world line, but each point on the world line will lie strictly inside the positive light cone at that point.

world. In order to develop the geometry of Minkowski space–time, we need to know how to measure distance in this universe. In mathematicians' terminology, we need a *metric*.

In a two-dimensional Euclidean universe, the normal metric is given by Pythagoras' theorem. The distance d of a point (x, y) from the origin is given by

$$d^2 = x^2 + y^2.$$

Likewise, in three-dimensional Euclidean space, the distance d of a point (x, y, z) from the origin is given by

$$d^2 = x^2 + y^2 + z^2,$$

and analogously in higher-dimensional Euclidean spaces. The metrics derived from Pythagoras' theorem in this way are called the *Pythagorean metrics*.

If we use the Pythagorean metric in Minkowski space–time, the distance d of a point (ct, x, y, z) from the origin is given by

$$d^2 = (ct)^2 + x^2 + y^2 + z^2.$$

Part of Minkowski's genius was that he realized that this was not the right metric for relativity theory, and discovered a metric that worked. The *Minkowski distance* d_M of a point (ct, x, y, z) from the origin is given by

$$d_M^2 = (ct)^2 - x^2 - y^2 - z^2.$$

To anyone familiar with Euclidean geometry, at first encounter, this definition of the Minkowski metric seems very strange. By rewriting the definition as

$$d_M^2 = (ct)^2 - (x^2 + y^2 + z^2),$$

we can see that the metric in the spatial coordinates is Euclidean. The puzzling feature is the minus sign, which forces us to treat the time coordinate very differently from the spatial coordinates. Let's see what we can make of this strange definition.

For any point (ct, x, y, z) in the surface of the light cone, we have $d_M = 0$. We can interpret this as saying that an event occurring at any point on the light cone is simultaneous to an event occurring at the origin.

If the point (ct, x, y, z) lies within the interior of the light cone, then $(ct)^2 > x^2 + y^2 + z^2$, and hence $d_M > 0$. In this case, we can interpret d_M as the time interval between an event at O and an event at (ct, x, y, z), measured by a clock whose world line is the straight-line segment from O to (ct, x, y, z). This is sometimes referred to as the *proper time* from O to (ct, x, y, z).

For a point (ct, x, y, z) that lies outside the light cone, $(ct)^2 < x^2 + y^2 + z^2$, and in this case d_M is an imaginary number. We have already observed that points outside the light cone do not exist as far as an observer at O is concerned (in the strong sense that they never have existed nor ever will exist from that observer's standpoint).

The Minkowski distance d_M between two points $P = (ct, x, y, z)$ and $Q = (ct', x', y', z')$ is given by

$$d_M{}^2 = (ct - ct')^2 - (x - x')^2 - (y - y')^2 - (z - z')^2.$$

This gives a real value for d_M, provided P is in the light cone of Q, and vice versa. In that case, d_M measures the time between an event at P and an event at Q, measured by a clock whose world line is the straight-line segment PQ, as illustrated in Figure 8.9—the proper time from P to Q. In the case in which the two events occur at the same location in physical space, so that $x = x'$, $y = y'$, $z = z'$, the above equation reduces to

$$d_M{}^2 = (ct - ct')^2,$$

and so $d_M = c|t - t'|$, the product of c and the actual elapsed time. In other words, Minkowski time at any particular physical location is the same as ordinary time at that location.

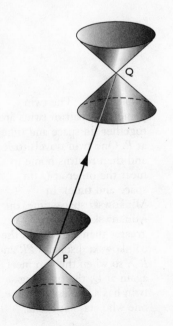

FIGURE **8.9** The world line of a massive particle traveling from point P to point Q in Minkowski space–time is a straight line that lies strictly within the positive light cone at P and the negative light cone at Q.

One curious feature of the Minkowski metric is that, in contrast to Euclidean geometry, in which the straight line from a point P to a point Q is the path from P to Q having minimal length, in Minkowski geometry, the straight, unaccelerated world line from a point P to a point Q is the path from P to Q for which the Minkowski distance is maximal (i.e., for which the proper time interval is greatest). This is the basis of the so-called twin paradox, illustrated in Figure 8.10.

Imagine a pair of twins at point P. The first twin, Homer, stays at home, on Earth. His sister, Rockette, goes on a round-trip to a distant planet R at a speed close to the speed of light. The twins are reunited at Q when Rockette completes her journey. When Rockette returns, she is younger than Homer. Less time has elapsed for Rockette than for Homer. This follows directly from the fact that the two-stage path from P to R to Q represents a shorter world line than the straight-line path from P to Q. (The latter has the maximal length of all world lines from P to Q, remember.) The path PRQ represents the passage of time for Rockette; the path PQ represents the passage of time for Homer.

It should be said that, for all its strangeness, there is considerable evidence to support the Minkowski metric for measuring time. Some of that evidence comes from measurements of the decay of particles in the cosmic radiation found in the earth's upper atmosphere. Other evidence comes from precise measurements of time in aircraft, and of the motion of particles in high-energy accelerators.

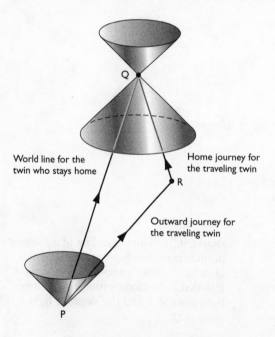

World line for the twin who stays home

Home journey for the traveling twin

R

Outward journey for the traveling twin

P

FIGURE **8.10** The twin paradox. A pair of twins are together (in space and time) at P. One twin travels to R and then returns home to meet the other at Q (in space and time). In Minkowski space–time, the Minkowski distance PQ is greater than the sum of the Minkowski distances PR and RQ, so when the two meet again at Q, the traveling twin has aged less than the one who stays at home.

In the case of the twin paradox, in order to obtain a noticeable age difference, the distance traveled by one of the twins has to be large, and the speed of travel most be close to the speed of light. Thus, the twin paradox is essentially a thought experiment. Where the distances involved are more realistic—say, a manned space flight to the moon—the time disparity is not noticeable. This is because of the magnitude of c: the term $(ct - ct')^2$ is very much larger than the term $(x - x')^2 + (y - y')^2 + (z - z')^2$ in the expression for the Minkowski metric. Under such a circumstance, the Minkowski distance d_M is approximately equal to $c|t - t'|$. That is to say, Minkowski time is approximately equal to ordinary time. It is only at very large physical distances, on an astronomical scale, that spatial distance affects time to a noticeable degree.

Patterns of gravity

To incorporate gravity into the geometric picture, Einstein replaced Minkowski space–time with a curved space–time manifold. We have met manifolds in Chapter 6; those we encountered there were what we might call *spatial* manifolds.

The general idea, you may recall, is that an n-dimensional (spatial) manifold is a structure that, in the immediate vicinity of any point, looks just like n-dimensional Euclidean space. For example, the (hollow) sphere and the (hollow) torus are both two-dimensional manifolds. In the immediate vicinity of any point, either surface looks just like the Euclidean plane. But the two surfaces have global structures that make them different from each other overall (and different from the Euclidean plane). They curve in different ways, and as a result they are distinct manifolds. Both of these manifolds are *smooth*—the differentiation structures that can be attached to each point fit together in a seamless fashion.

Realizing that gravity could be regarded as a manifestation of a curving of space–time, Einstein took the underlying mathematical structure of the universe to be a smooth four-dimensional *space–time manifold*. This would mean that, in the immediate vicinity of any point, the universe looks just like Minkowski space–time. To construct such an object mathematically, Einstein started with a smooth, four-dimensional manifold, $\mathcal{M}$. He then defined a metric on $\mathcal{M}$ that, in the immediate vicinity of any point, can be approximated by the Minkowski metric. The approximating Minkowski metric at any point gives rise to a light cone at that point. The new metric Einstein defined can be used to describe the passing of time, the propagation of light beams, and the motion (accelerated or unaccelerated) of particles, planets, and so forth.

Since Einstein was trying to describe gravity, he had to connect the underlying geometry of the universe—represented by the manifold $\mathcal{M}$—with the distribution of matter in the universe. He represented the latter, the distribution of mass, by means of a mathematical structure called an *energy–momentum tensor*. It would require far too great a digression to describe this somewhat complicated object. It corresponds to—and in some ways can be thought of as a generalization of—the density of mass in Newton's (far simpler) gravitational theory. A key step in Newton's theory was the formulation of a differential equation (Poisson's equation) connecting the density of mass with what he called a *gravity potential*, which corresponds to Einstein's metric. Analogously, Einstein wrote down an equation (Einstein's equation) that connects the energy–momentum tensor with the curvature derived from the metric.

According to Einstein's equation, the presence of matter introduces a metrical curvature in space–time, and this curvature affects other objects in the universe. Newton's idea of a physical *force* of gravity is replaced by the geometric concept of *curvature* of space–time.

Another consequence of Einstein's theory is that mass and energy are interchangeable, as related by his famous equation

$$E = mc^2.$$

Is Einstein's theory right? Is space–time really curved? Well, strictly speaking, it is meaningless to ask whether a scientific theory is right. All one can ask is that it provide better agreement with the observations—and more accurate predictions—than any other theory. In this respect, Einstein's theory of general relativity is more accurate than Newton's earlier theory. In recent years, very accurate measuring techniques have been used to detect small deviations from flat Minkowski space–time caused by a gravitational field. For instance, in 1960, Robert V. Pound and Glen A. Rebka measured the relationship between the speed of a clock at the top and at the bottom of a 22.6 meter high tower. The answer was not 1, as predicted by Newton's theory, but 1.000 000 000 000 0025. This is exactly the ratio predicted by general relativity. (Since clock speeds measure the geometry of space–time, we can take the Pound–Rebka result as a direct measure of the deviation of real space–time from a flat Minkowski space–time.)

What's the matter?

While the theory of general relativity describes the geometric structure of the universe, and shows how matter affects, and is affected by, that geometric structure, it does not answer the question: what exactly *is* matter? To answer that question, physicists had to turn to another theory: *quantum theory*.

By the early 1920s, the standard view of matter was that it was made up of atoms—tiny solar system-like entities, each consisting of a heavy nucleus (the 'sun' of the atom) around which orbited one or more much lighter electrons (the 'planets'). The nucleus of the atom was itself supposed to be made up of two kinds of fundamental particles: protons and neutrons. Each proton carried a positive charge and each electron a negative charge, and it was the electromagnetic attraction between these charges that provided the 'gravitational force' holding the electrons in their orbits around the nucleus. (This picture is still a useful one, even though it is now known to be overly simple.)

But what exactly are these fundamental particles—the electrons, the protons, and the neutrons—of which atoms are built? That was the question facing the physicists Niels Bohr, Werner Heisenberg, Erwin Schrödinger, and others around 1920. In order to take into account some puzzling experimental results, the answer they provided—quantum theory—incorporated an element of randomness in a fundamental way, a move that infuriated Einstein, who at one time remarked, "God does not play dice with the universe." (One puzzling phenomenon that quantum theory explained was that light seemed to behave like a continuous wave in some circumstances and a stream of discrete particles in others.)

In quantum theory, a particle, or any other physical object, is described by a probability distribution, which expresses the particle's or object's tendency to be in a given state. In mathematical terms, a particle is described by a wave function ψ, which with each point x in Einstein's space–time manifold $\mathcal{M}$ associates a vector $\psi(x)$, whose magnitude represents the amplitude of the oscillation and whose direction gives its phase. The square of the length of $\psi(x)$ gives the probability that the particle is close to the point x in $\mathcal{M}$.

When quantum theory is taken as the basic framework, the classic picture of material particles occupying a surrounding space disappears. In its place there is a *quantum field*, a fundamental continuous medium present everywhere in space. Particles are just local densities—concentrations of energy—in the quantum field.

At this point let us pause for a moment and look again at that seemingly simple solar-system picture of the atom portrayed a moment ago. We know what keeps the electrons in their orbits around the nucleus: the electromagnetic force that causes oppositely charged particles to attract each other. But what holds the protons together in the nucleus? After all, like charges repel. Why doesn't the nucleus simply explode?

There must be some force—and a massive one at that—that holds the nucleus together. Physicists call it the *strong nuclear force*, or the *strong interaction*. The strong interaction must be strong indeed in order to hold together the protons in a single nucleus. On the other hand, it must act only over a very short distance, of the order of the nucleus itself, since it does not pull together the protons in the nuclei of two distinct atoms—if it did, we ourselves would implode, along with everything else in the universe.

The strong nuclear force is assumed to be a fundamental force of nature, along with gravity and the electromagnetic force. In fact, physicists believe there is a fourth fundamental force. In order to explain other nuclear phenomena having to do with nuclear decay, they have proposed a second kind of nuclear force: the *weak nuclear force*, or the *weak interaction*.

At this point, the skeptic might be inclined to think that there are still further natural forces waiting to be discovered. While this possibility cannot be completely ruled out, physicists don't think so. They believe that gravity, the electromagnetic force, the strong nuclear force, and the weak nuclear force are all there are.

Assuming that the physicists are right, and there are just four fundamental forces, then a complete theory of matter will have to incorporate all four. Physicists have struggled hard—hitherto without complete success—to find a single mathematical theory that encompasses them all. The key mathematical idea that they think will eventually lead them to such a theory is symmetry.

We first encountered the concept of symmetry in Chapter 5. Recall that an object is said to be symmetrical if some transformation leaves it looking exactly as it did in the first place. We observed that geometry can be regarded as the study of those properties of objects or figures that do not change when the objects or figures are moved around. We also noticed that the symmetries of an object form a mathematical structure known as a group—the symmetry group for the object in question.

Early in the twentieth century, physicists began to realize that many of their conservation laws (such as the law of conservation of electric charge) arise from symmetries in the structure of the universe. For example, many physical properties are invariant under translation and rotation. The results of an experiment do not depend on where the laboratory is situated or the direction the equipment faces. This invariance implies the classic physical laws of conservation of momentum and angular momentum.

In fact, the German mathematician Emmy Noether proved that every conservation law can be regarded as the result of some symmetry.

Thus, every conservation law has an associated symmetry group. For example, the classic law of conservation of electric charge has an associated symmetry group. So too do the more modern quantum physicist's laws of conservation of 'strangeness', of 'spin', and so forth.

In 1918, Hermann Weyl set out to unify the theories of general relativity and electromagnetism. His starting point was the observation that Maxwell's equations are invariant (i.e., symmetrical) with respect to a change of scale. He tried to make use of this fact by regarding the electromagnetic field as the distortion of the relativistic length that arises when you travel around a closed path (such as a circle). To do this, he had to assign a symmetry group to each point of four-dimensional space–time. The name Weyl gave to his new approach was *gauge theory*—a gauge being a kind of measuring device, of course. The group assigned to each point in space–time is called the *gauge group*.

Weyl's initial approach did not completely work. With the emergence of quantum theory, with its emphasis on the wave function, it became clear what the problem was. It wasn't scale that was important in Maxwell's equations; the crucial notion was phase. Weyl had been working with the wrong symmetry group! With his emphasis on scale, the gauge group he had been working with was the group of positive real numbers (with multiplication as the group operation). With the change in focus from scale to phase, the corresponding gauge group was the rotations of a circle.

Once he had found the right group, Weyl quickly developed his new electromagnetic theory. It is known as *quantum electrodynamics*, or QED for short, and it has been a hotbed of activity for mathematicians and physicists ever since.

With the stage set by Weyl, physicists set about trying to find a *grand unified theory* (of matter and the universe, incorporating the four fundamental forces), using gauge theory as their principal tool. The general idea was to find a gauge group that would enable them to capture the four different kinds of forces.

In the 1970s, Abdus Salam, Sheldon Glashow, and Steven Weinberg succeeded in unifying electrodynamics and the weak interactions in a gauge theory having as its gauge group a group known as U(2), a group of rotations in two-dimensional complex number space. The three were awarded the Nobel Prize for this achievement.

The next step was taken by Glashow and Howard Georgi a decade later. Using a larger gauge group known as SU(3) × U(2), they managed to incorporate the strong interactions as well.

The final step, incorporating gravity, has so far proved elusive, and remains something of a physicists' Holy Grail. In recent years, physicists have looked to other theories, more general than gauge theory, to complete this last step. By far the most common is *string theory*, of which the leading developer is Edward Witten.

In string theory, the fundamental objects are no longer particles moving along curves (world lines) in a space–time manifold, but tiny open or closed strings, each sweeping out a two-dimensional surface called a *world surface*. To make this work, physicists have had to learn to adapt to working in a space–time manifold of at least ten dimensions. If there are fewer dimensions, there is not enough freedom for the world surfaces to be swept out properly.

And if you think ten dimensions is a lot, there are versions of string theory that require twenty-six dimensions. Though their ultimate aim is to understand the familiar three-dimensional world we live in, physicists have thus been led into more and more abstract mathematical universes. Whether or not our universe is accurately described as a 'mathematical universe', the only way we have to try to understand that universe is *through* mathematics.

Vice versa

Everyone knows that physicists use lots of mathematics. Newton developed his own—the calculus—and used it to study the universe. Einstein used non-Euclidean geometry and manifold theory to develop his theory of relativity. Weyl developed gauge theory to support his QED theory. And so forth. Applying mathematics to physics is a time-honored practice.

Far less common is going in the other direction: taking ideas and methods from physics and applying them to obtain new discoveries in mathematics. But that is exactly what has been happening over the last fifteen years or so. The origins of this dramatic shift in who gives what to whom go back to Weyl's development of gauge theory in 1929.

Because the group Weyl used—the rotations of a circle—was abelian (i.e., commutative; see Chapter 5), the resulting theory was called *abelian gauge theory*. In the 1950s, some physicists, initially motivated by pure speculation, began to investigate what would happen if they changed the group, say, to the symmetries of a sphere, or of other, higher-dimensional objects. The resulting theory would be a *nonabelian gauge theory*.

Two of those curiosity-driven physicists who started to investigate nonabelian gauge theories were Chen-Ning Yang and Robert Mills. In 1954, they formulated a collection of fundamental equations analogous to Maxwell's equations. Just as Maxwell's equations describe an electromagnetic field and are associated with an abelian group, the Yang–Mills equations also describe a field, but in this case the field provides a means of describing particle interactions. Moreover, the associated group is nonabelian.

At this point, some mathematicians started to take notice of these startling new developments in physics. Though physicists were using the Yang–Mills equations within the framework of quantum theory, the equations also had a classic (i.e., nonquantum) version, with ordinary four-dimensional space–time as the underlying manifold. It was this version of the equations that the mathematicians concentrated on at first.

The mathematicians were able to solve a special case of the Yang–Mills equations (the so-called 'self-dual' case), calling their solutions *instantons*. Though of some interest to the physicists, these new mathematical results were still very much a pure mathematician's game. But then along came a young British mathematician named Simon Donaldson, whom we have encountered once already in Chapter 6.

Donaldson applied instantons and the self-dual Yang–Mills equations to general four-dimensional manifolds, thereby opening the door on a dramatic and startling new mathematical vista, starting with the surprising discovery of the nonstandard differentiation structure for four-dimensional Euclidean space mentioned in Chapter 6.

The aspect of Donaldson's work I want to emphasize here is that it involved a reversal of the usual flow of ideas. As we observed a moment ago, mathematicians and physicists are used to the notion that physics makes use of mathematics. But with Donaldson's work, the flow was in the opposite direction, and it heralded a new era in which geometry and physics would become intertwined to a degree that no one had anticipated.

One important consequence of Donaldson's theory was that it provided a means of generating invariants for 4-manifolds. Suddenly, mathematicians had a way to approach the classification problem for 4-manifolds that had so far resisted all their attempts. But the going was decidedly tough: a lot of work was required to coax Donaldson's theory into turning out the desired invariants.

Donaldson's former teacher Michael Atiyah was sure that the solution to the difficulty—and the solution to many difficulties the physicists were facing—was, essentially, to reunite mathematics and physics. Inspired by Atiyah's suggestion, in 1988 Princeton physicist Edward Witten managed to interpret Donaldson's theory as a quantum Yang–Mills theory. Apart from making Donaldson's work intelligible to the physics community, however, little came of Witten's work until 1993. Then physicist Nathan Seiberg entered the picture, and things changed dramatically.

Seiberg had been looking at supersymmetry, a theory introduced in the 1970s, which posits an overarching symmetry between the two main classes of elementary particles, the fermions (which include electrons and another class of particles called quarks) and the bosons (which include photons—particles of light—and still another collection of particles called gluons). He had developed methods of handling the difficult problems of quantum gauge theories for the supersymmetrical case.

In 1993, Seiberg and Witten embarked on a joint investigation of the case related to Donaldson's theory. In what Witten subsequently described as "one of the most surprising experiences of my life," the two of them found a new pair of equations to replace the ones coming from Donaldson's theory. This marked major progress both in physics and along the mathematical road to the long-sought-after classification of 4-manifolds.

The underlying distinction between Donaldson's theory and the new Seiberg–Witten theory that is the key to the power of the latter is *compactness*. Compactness is a technical property of topological spaces. Intuitively, it means that if each point of the space gives you information about its immediate neighborhood, you can glean all the combined information you need from just a finite number of the points. The space associated with Donaldson's theory is not compact; the Seiberg–Witten one is. Compactness makes all the difference.

The Seiberg–Witten discovery had immediate consequences not only for physics but also for mathematics. Though his main interest was in the physics, Witten worked out the implications of the new discovery for 4-manifolds. He conjectured that the new equations would produce invariants identical to those coming from Donaldson's theory.

Witten's conjecture was taken up by Harvard mathematician Clifford Taubes and others, and pretty soon it was realized that something dramatic had happened. It is not yet known whether the invariants the new equations give rise to are exactly the same as Donaldson's invariants. However, they are at least as good and, crucially, are much, much easier to deal with. Using the new equations, mathematicians were able to redo in a few weeks, and with far less effort, what had taken years using the old methods, and to solve problems they had not been able to crack before. At last, mathematicians sensed real progress toward an understanding of 4-manifolds. After receiving three hundred years of excellent service from mathematics, physics had finally found a way to repay the debt.

And there I'll stop. But mathematicians go on, continuing what will be a never-ending search to understand the hidden patterns of the universe and the life within it.

Postscript

There is more. A great deal more. The general themes outlined in the previous chapters cover just one small portion of present-day mathematics. There was little or no mention of computation theory, computational complexity theory, numerical analysis, approximation theory, dynamical systems theory, chaos theory, the theory of infinite numbers, game theory, the theory of voting systems, the theory of conflict, operations research, optimization theory, mathematical economics, the mathematics of finance, catastrophe theory, or weather forecasting. Little or no mention was made of applications of mathematics in engineering, in astronomy, in psychology, in biology, in chemistry, in ecology, and in aerospace science. Each one of these topics could have occupied a complete chapter in a book of this nature. So too could a number of other topics that I did not include in the above lists.

In writing any book, an author must make choices. In writing this book, I wanted to convey some sense of the nature of mathematics, both the mathematics of today and its evolution over the course of history. But I did not want to serve up a vast smorgasbord of topics, each one allotted a couple of pages. For all its many facets and its many connections to other disciplines and other walks of life, mathematics is very much a single whole. A mathematical study of any one phenomenon has many similarities to a mathematical study of any other. There is an initial simplification, in which the key concepts are identified and isolated. Then those key concepts are analyzed in greater and greater depth, as relevant patterns are discovered and investigated. There are attempts at axiomatization. The level of abstraction increases. Theorems are formulated and proved. Connections to other parts of mathematics are uncovered or suspected. The theory is generalized, leading to the discovery of further similarities to—and connections with—other areas of mathematics.

It is this overall structure of the field that I wanted to convey. The particular topics I chose are all central themes within mathematics, in that they are all included, to a greater or lesser extent, in many, if not most, college or university undergraduate degree programs in mathematics. In that sense, their selection was a natural one. But the fact is, I could have chosen any collection of seven or eight general areas and told the same story: That mathematics is the science of patterns, and those patterns can be found anywhere you care to look for them, in the physical universe, in the living world, or even in our own minds. And that mathematics serves us by making the invisible visible.

Stop the presses!

Mathematics marches on. Just as this volume was going to press, it was announced that Kepler's sphere packing problem (page 200) had been solved. After six years of effort, mathematician Thomas Hales of the University of Michigan managed to show that Kepler's conjecture was correct: The face-centered cubic lattice is inded the densest of all three-dimensional sphere packings.

Hales's proof was based on the work of a Hungarian mathematician, Laszlo Toth, who, in 1953, showed how to reduce the problem to a huge calculation involving many specific cases. That opened the door to solving the problem using a computer. In 1994, Hales worked out a five-step strategy for solving the problem along the lines Toth had suggested. Together with his graduate student Samuel Ferguson, Hales started work on his five-step program. In August 1998, Hales announced that he had succeeded, posting the entire proof on the World Wide Web at the address http://www.math.lsa.umich.edu/~hales/.

Hale's proof involves 250 pages of text and about 3 gigabytes of computer programs and data. Any mathematician who wants to follow Hale's argument must not only read the text, but also must download Hales's programs and run them.

In using a combination of traditional mathematical reasoning and massive computer calculations dealing with many hundreds of special cases, Hales's solution to Kepler's problem is reminiscent of the solution in 1976 of the four color problem by Kenneth Appel and Wolfgang Haken, discussed in Chapter 6.

Index